A Journal of the Plague Year

A Journal of the Plague Year

THE DIARY OF
THE BARCELONA TANNER
MIQUEL PARETS
1651

JAMES S. AMELANG
Translator and Editor

New York Oxford
OXFORD UNIVERSITY PRESS
1991

Oxford University Press

Oxford New York Toronto
Delhi Bombay Calcutta Madras Karachi
Petaling Jaya Singapore Hong Kong Tokyo
Nairobi Dar es Salaam Cape Town
Melbourne Auckland

and associated companies in
Berlin Ibadan

Published by Oxford University Press, Inc.,
200 Madison Avenue, New York, New York 10016

Oxford is a registered trademark of Oxford University Press

Library of Congress Cataloging-in-Publication Data
Parets, Miquel.
[Molts successos que han succeït dins Barcelona i molts altres
llocs de Catalunya, dignes de memòria. English. Selections]
A journal of the plague year : the diary of the Barcelona tanner
Miquel Parets, 1651 / James S. Amelang, translator and editor.
p. cm. Translation of selections from:
De molts successos que han succeït dins Barcelona i molts
altres llocs de Catalunya, dignes de memòria.
Includes bibliographical references. ISBN 0-19-506455-0
1. Plague—Spain—Barcelona—Early works to 1800. 2. Parets, Miquel—
Diaries. I. Amelang, James S., 1952- . II. Title.
RC178.S7P372513 1991 946'.72'052—dc20 90-33457

9 8 7 6 5 4 3 2 1

Printed in the United States of America
on acid-free paper

To Elena, Daniel, and David

Nor for my peace will I go far,
As wanderers do, that still do roam,
But make my strengths, such as they are,
Here in my bosom, and at home.

Ben Jonson

In any event, his notebooks can be read as a chronicle of those difficult times. But one is dealing with a very special sort of chronicle, one which seems deliberately devoted to recording the least significant details. . . . Amid the general confusion he decided, in effect, to write about those matters usually neglected by historians. One can certainly have second thoughts about this choice, and suspect a certain hardness of heart on his part. All the same, his notebooks can provide, for a chronicle of this period, a host of minor details which are nevertheless important, and whose very singularity prevents one from judging too quickly this rather strange person.

Albert Camus, *The Plague*

ACKNOWLEDGMENTS

It is a pleasure to acknowledge the assistance of numerous friends and colleagues in the preparation of this book.

An early draft of the introduction was presented as a paper at the First Spanish-Portuguese-Italian Conference on Historical Demography held in Barcelona in April 1987. I am indebted to the organizers of the panel on the Mediterranean plague of the mid-seventeenth century, Bruno Anatra and Vicente Pérez Moreda, for inviting a non-demographer to offer his reflections *in partibus infidelium*. I have since delivered other papers and talks on Parets and the plague at the University of Santiago de Compostela (Spain), the University of Florida, the Shelby Cullom Davis Center of the History Department at Princeton University, the Social History Seminar of the University of Chicago, the University of Maryland, and Southern Methodist University. That the final product bears scant relation to its original form owes much to the thoughtful criticism and commentary I received at all these institutions.

I have also benefitted from the suggestions and comments of other scholars who read drafts of the introductory essay, including Judith Brown, Tom Cohen, Jim Farr, Gary McDonogh, Ed Muir, and Lou Rose. Juan Eloy Gelabert, Xavier Gil, Rosemarijn Hoefte, John Martin, Laurie Nussdorfer, Alessandro Pastore and Richard Palmer kindly supplied bibliographic references, often upon rather short notice. Xavier Torres, who co-edited the Catalan edition of the Parets manuscript, also provided invaluable assistance in the preparation of the English version of this text. I am particularly indebted to Rob Bartlett, Peter Sahlins, and my copy editor Linda Grossman for help with the final version of the introduction. Both Bill Christian and Richard Kagan read the entire manuscript and offered numerous suggestions for its improvement. The kindness of all these scholars does not, however, exempt myself from responsibility for any errors or infelicities it may contain.

Funding for research on a larger project involving Miquel Parets and artisan culture in early modern Barcelona has been provided by the U.S.–Spanish Joint Committee for Cultural and Education Cooperation, the Tinker Foundation (through a grant administered by the Center for Latin American Studies, University of Florida), and the History Department of the University of Florida. I also am grateful to Sr. Jordi Torra, of the Reserve Section of the University of Barcelona Library, for facilitating the consultation and reproduction of Parets's original manuscript. Finally, both the editor and the publisher would like to acknowledge EUMO Editorial, of Vic, Spain, and especially Joaquim Albareda, for their generous cooperation in the preparation of this text.

CONTENTS

BARCELONA IN 1651
(FORTIFICATIONS NOT DEPICTED)
LEGEND*
Religious buildings:
Parish churches
Monasteries and convents
Chapels
Other religious buildings
Secular buildings
Numbered streets:
-1- Passageway of the Prison
-2- Knifemakers' Street
-3- Moneychangers' Street
-4- Street of Hell
-5- Booksellers' Street
-6- St. James' Square
-7- King's Square
*See Appendix IV for Catalan names of streets and plazas.
St. Severus' Tower
Nazareth
(Third Pesthouse)
BUTCHERS' STREET
NAZARETH STREET
Seminary
Angels Convent
(Angels nous)
Carmen
CARMEN STREET
Royal Hospital
St. Anthony's Gate
HOSPITAL STREET
ST. PAUL'S STREET
ST. PAUL'S QUARTER
St. Paul's Tower
St. Monica
Naval Yards
(Dressanes)
St. Madrona Bastion
0
meters
400
(appx. ¼ mile)
Angel Gate
University
St. Anne's
ST. ANNE'S ST.
ST. ANNE'S SQ.
Mount Zion
RAMBLA
St. Mary's of the Pine
BOCARIA
NEW STREET
St. Francis de Paul
ST. JOHN'S WAY
Magdalenes
NEW SQ.
Bishop's Palace
Cathedral
Inquisition
(Royal Palace)
Meat Shop
(Carnisseria)
St. Agatha
Prison
Diputació
St. James'
CALL
City Hall
REGOMIR
St. Justus'
BROAD STREET
Sea Wall
St. Catherine's
Parets House
Marcus
BORIA
St. Augustine
New Gate
CURRIERS' SQ.
(Blanqueria)
LLANSANA ST.
FUSINA
MIRRORMAKERS' ST.
HATTERS' ST.
BORN
St. Mary's by the Sea
STREET OF THE HOLY SPIRIT
St. Daniel's Gate
East Bastion
AUCTION SQ.
St. Sebastian
Sea Gate
Customs House
South Bastion
MEDITERRNEAN SEA

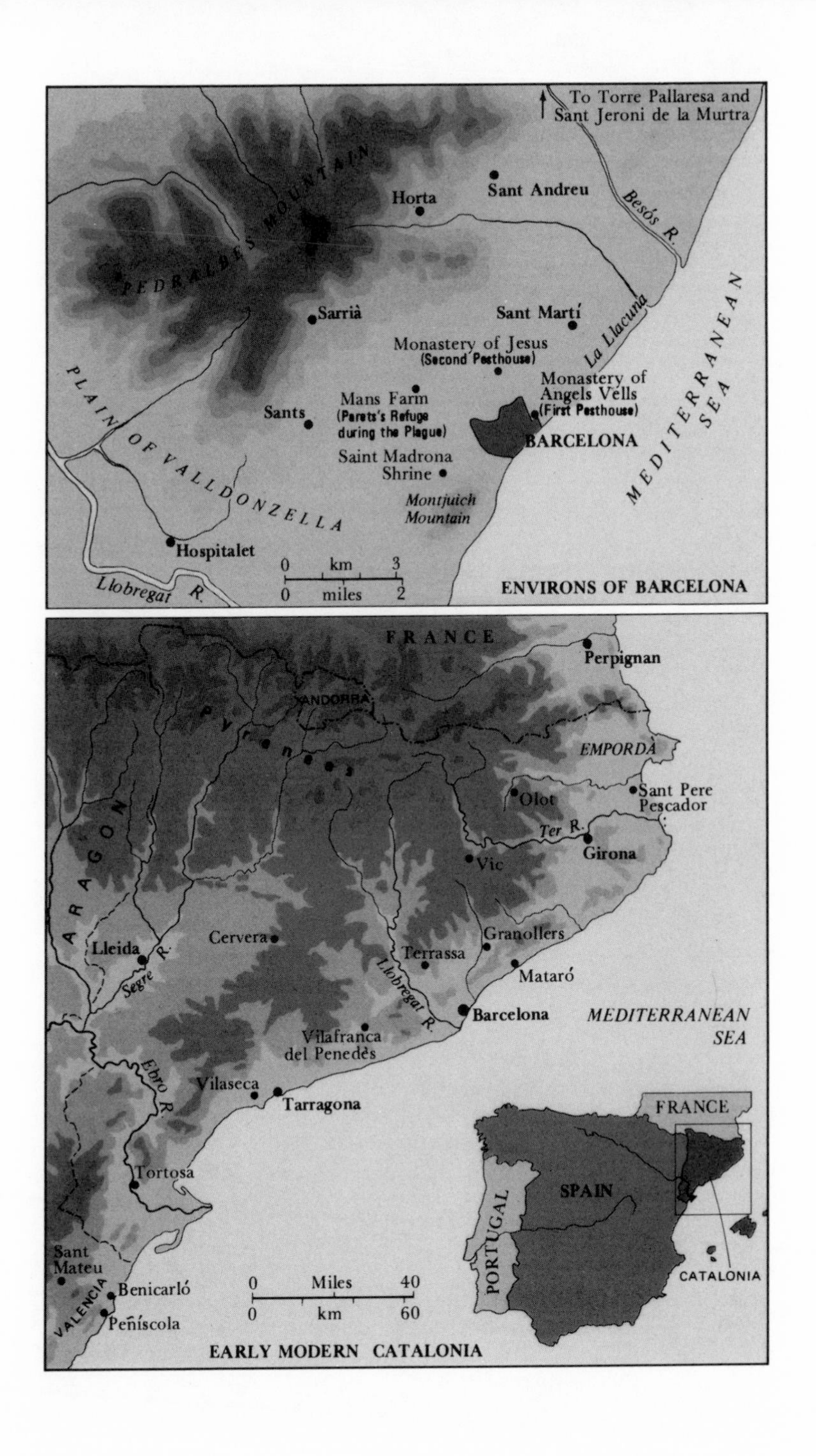

To Torre Pallaresa and Sant Jeroni de la Murtra
PEDRALBES MOUNTAIN
Horta
Sant Andreu
Besós R.
Sarrià
Sant Martí
La Llacuna
Monastery of Jesus (Second Pesthouse)
Mans Farm (Parets's Refuge during the Plague)
Monastery of Angels Vells (First Pesthouse)
Sants
BARCELONA
MEDITERRANEAN SEA
PLAIN OF VALLDONZELLA
Saint Madrona Shrine
Montjuich Mountain
Hospitalet
Llobregat R.
0 km 3
0 miles 2
ENVIRONS OF BARCELONA
FRANCE
Perpignan
ANDORRA
Pyrenees
EMPORDÀ
Sant Pere Pescador
Olot
Ter R.
Girona
Vic
ARAGON
Granollers
Cervera
Lleida
Terrassa
Segre R.
Llobregat R.
Mataró
Barcelona
MEDITERRANEAN SEA
Vilafranca del Penedès
Ebro R.
Vilaseca
Tarragona
FRANCE
Tortosa
SPAIN
PORTUGAL
Sant Mateu
CATALONIA
Benicarló
VALENCIA
Peñíscola
0 Miles 40
0 km 60
EARLY MODERN CATALONIA

A Journal of
the Plague Year

Introduction: Popular Narrative and the Plague

In the summer of 1665, an anonymous saddler began a diary which he hoped would convey to future readers the terror and devastation caused by the epidemic then raging through London. A natural sense of curiosity impelled him, he confessed, to hazard long walks through the plague-ridden city. When not braving the streets, however, he gave over his time to "writing down my memorandums of what occurred to me every day, and out of which, afterwards, I took most of this work, as it relates to my observations without doors. What I wrote of my private meditations I reserve for private use, and desire it may not be made publick on any account whatever."[1] These entries formed the nucleus of what eventually became the most famous early modern plague narrative, the *Journal of the Plague Year*, published by Daniel Defoe in 1722.

I have borrowed Defoe's title for this book, which centers around a very similar document—the chronicle of the Barcelona plague of 1651 written by the master tanner Miquel Parets. Parets's account differs from Defoe's text in at least three respects. First, it is probable that the nameless saddler of London never existed, and that his narrative was a fictional creation—in the form of a thoroughly convincing historical reconstruction—written by Defoe himself.[2] The Parets chronicle is, on the other hand, authentic. In fact, it is the most extensive surviving popular account of the experience of plague in European history. Second, Defoe (born around 1660) did not himself live through England's last great outbreak of plague as an adult. The Barcelona craftsman, to his misfortune, was a direct eyewitness and survivor of the epidemic which desolated his city and destroyed his family. Finally, unlike his English counterpart, the Catalan artisan reserved none of his "private meditations for private use." As will be seen, his is a naked tale of anguish and grief—a deeply revealing chronicle of a city, and a man, under siege.

"Toward the end of July in a house on Saint Vincent's street in Valencia . . . many died, and in so short a time that it could not fail to be noticed. . . . By the beginning of August the disease had begun to spread throughout the city." Thus began Friar Francisco Gavaldá's chilling account of the plague's entry into Spain in 1647.[3] Within weeks the epidemic spread outward from its origins in and near the city of Valencia. After devastating much of Andalusia, it expanded northward into Aragon and Catalonia, and from there to Mallorca and Sardinia. While four years later it had all but disappeared within the peninsula, its resurgence in western Italy in 1655 wrote a tragic coda to what one scholar has labelled "the greatest catastrophe to strike Spain in the early modern period."[4]

Despite extensive preventive measures ordered by the city councillors, the disease reached Barcelona shortly after New Year's Day 1651.[5] The initial (and all too familiar) reluctance of physicians to label openly the disease as plague inspired no more confidence among local citizens than municipal officials' attempts to keep secret the establishment of extramural pesthouses. Within weeks thousands fled the city for the neighboring countryside. Meanwhile, those who remained attempted to stem the epidemic's advance by applying a tandem of practical measures (including closing most of the city's gates) and metaphysical remedies (such as rogative processions).[6] Plague orders and patron saints failed to halt the spread of the disease; by August, when a *Te Deum* was sung to celebrate its temporary departure, perhaps as many as one third of Barcelona's 45,000 inhabitants had died.

The human and material impoverishment of the capital of Catalonia in 1651–1652 provides an ironic contrast to the wealth of documents testifying to its sufferings. During these years of famine, plague, and war, the city's notaries dutifully continued their meticulous annotations in the municipal council's minute-books, fiscal accounts, and registry of ordinances. The result of their diligence is a rich and comprehensive documentary record, especially useful for reconstructing public and institutional reactions to collective disaster.[7] The routine impersonality of these sources, however, prevents our penetrating the private world of sentiments, fears, and other emotive responses to epidemics. Hence the value of intimate records providing testimony to the direct experience of plague.

The notions of testimony and experience suggest a more general question: how did the survivors of catastrophes like the plague recall their ordeal? How did plague, in other words, construct historical memory, both individual and collective? Memoirs of (and memorials to) plagues took many forms. Some were physical, ranging from the miniature, like the candle the pinmakers of late medieval York kept lit in ongoing remembrance of the "great pestilence" of 1349,[8] to the monumental, such as the *Pestsäule*, or baroque column, erected in Vienna's Graben square to commemorate the victims of the epidemic of 1682. Other material reminders of the visitation of plague were votive churches, like the Redentore and Santa Maria della Salute in Venice, constructed to fulfill civic vows made during the epidemics of 1576 and 1630, respectively. Venice also provides an example of another type of plague memorial; its Scuola di San Rocco, a devotional confraternity famed for its sumptuous building and lavish frescoes by Tintoretto, was also a plague foundation (dating to 1478). Throughout Europe, pestilence proved a popular subject in painting, drawing, and sculpture, ranging from thousands of portraits and statues of the plague saints Sebastian and Roche, to direct historical depictions, like Micco Spadara's terrifying canvas of the marketplace in Naples during the contagion of 1656.[9]

Most of the testimonies of plague survivors were written, however, and it is to the documentary record that one must turn in order to reconstruct the experience of epidemics. Personal narratives of the plague are both numerous and varied in form, as a wide range of documents incorporated memoirs of the disease (see Appendix II). Accounts by clerics constitute one important source.[10] These include notations within parish registers, special "books of the dead," which were kept in times of extraordinary mortality, correspondence, and, of course, clergymen's diaries. A second category involves the plague narratives found within official government correspondence, such as reports by provincial officers to central administrative bodies like the

Privy Council in England or the Council of Castile in Spain.[11] Medical treatises (as well as autopsies and forensic descriptions) also often contained accounts of direct experience with plague.[12] A good example from early modern Spain is Juan Tomás Porcell's influential work on the Saragossa epidemic of 1565, a text which strongly urged contemporary physicians to learn from firsthand contact with the contagion.[13]

Authors of fiction and devotional works also employed various literary forms, both popular and learned, to comment upon or recreate the personal experience of plague. The loss of family members and friends to "contagion" inspired poems by Petrarch and Ben Jonson, plays by Thomas Dekker, and sermons by John Donne. Epidemics loomed large as a theme in the hundreds of chapbooks and broadsheets anonymous authors published during the sixteenth and seventeenth centuries.[14] More generally, references to plague were adopted as metaphors by numerous early modern writers regardless of whether or not they had ever come into direct contact with the disease. It was not rare, for example, to find sixteenth-century Protestant and Catholic apologists referring to each other as "pestilential."[15] Along different lines, Shakespeare often pressed plague into service as an experientially neutral simile for personal and social relations; hence his obsessive use of the term in *Twelfth Night* to denote the sudden "infection" of love. This point—that references to epidemics served a broad spectrum of purposes, ranging from the personalizing to the polemical, within diverse early modern literary contexts—should nevertheless not be pressed too far. More striking, in fact, is the general *lack* of fictional depictions and evocations of plague in medieval and early modern European literature, and this seems especially true of Spain.[16]

Arguably the most important—certainly the most revealing—personal "plague texts" can be found in early modern diaries and autobiographies. On the continent, well-known diarists like Pierre de l'Étoile, Claude Haton, and Thomas Platter left harrowing first-person accounts of the disease. Across the Channel, the correspondence and journals of John Chamberlain, John Evelyn, Richard Baxter, and Samuel Pepys also recorded their contact with plague. Apart from the published memoirs of Gavaldá, sustained personal commentary on the plague in early modern Iberia can be found in journals from: Lisbon (1569); Galicia (late sixteenth century); Málaga, Seville, Girona, and Saragossa (mid-seventeenth century); and Cádiz (1681). More fleeting mention of contagion appears in other early modern Spanish personal literature. The Catalan Jesuit Pere Gil wrote a brief but fascinating recollection of the 1589 Barcelona epidemic; his Andalusian counterpart Pedro de León, who survived catching the plague in 1601, unfortunately made only the most cursory mention of his experience. Brief mentions of the Valladolid epidemic of 1599–1600 can be found in a contemporary treatise on political reform and in the correspondence of an Italian banker resident in that city, while both the notary Pere Pascual and the surgeon Jeroni Cros recounted their flight from the Perpignan plague of 1631 in their diaries.[17] There survive, in fact, several personal accounts of plague in Catalonia from 1650 to 1653, ranging from narratives written by a Barcelona patrician, a noble from the nearby city of Girona, and a notary from the town of Vic, to the artisan's account reproduced later.[18]

Different generations of historians have put these plague texts to varying uses.[19] Chronicles have long served as sources for the massive body of historical studies of epidemics in medieval and early modern Europe. In fact, plague narratives rank among the oldest forms of historical writings, the earliest being Thucydides' account of the "plague of Athens" of 430 BC.[20] Chronicles were also among the sources most frequently used by the first modern students of the disease, who in the nineteenth and early twentieth centuries produced numerous local histories of epidemics. Succeeding generations of historians, however, reacted against the impressionistic and descriptive nature of these accounts by turning to more objective themes and sources, placing special emphasis on the demographic and economic impact of the plague.[21]

Recent studies of mortality and attitudes toward death in the early modern era reflect the dramatic resurgence of interest in the more subjective side of human experience.[22] The present emphasis on the psychological and cultural dimensions of the experience of plague and other collective disasters has fostered a return to personal chronicles, autobiographies, and diaries as forms of historical evidence. Documents like the Parets memoirs—judged by many earlier scholars to be of "little or no interest"[23] for anything but the information they provided on contemporary politics and diplomacy—are now seen as irreplaceable sources by the historian seeking to penetrate beyond the surface to the inner lives, thoughts, and emotions of citizens of bygone times. Few of these documents, however, derive from circles outside a literate elite composed of notables, physicians, clergy, amateur historians, and the professionally curious. There are, to be sure, exceptions to this rule. Thus far I have been able to locate over a dozen popular plague accounts, from which some excerpts are in Appendix I.[24] These range from the memoirs of a sixteenth-century German goldsmith to diaries by printers from Basel and Málaga, a carpenter and unidentified *popolano* (craftsman) from Milan, an Aragonese surgeon, a Catalan peasant, a shopkeeper from Seville, a Bolognese glassmaker, and a memoir by a journeyman weaver in the northern French city of Lille. Further research would doubtless lengthen this list. However, as a rule popular plague narratives come few and far between. The scarcity of plague accounts by artisans and peasants thus renders all the more valuable the extensive treatment accorded to the outbreak of 1651 by the Barcelona craftsman.

Few details are known of Parets's life.[25] Born in 1610, he was named after his father, also a tanner. At some point before the latter's death in 1631, Parets was commissioned a master and admitted to the tanners' guild. He later married—three times—and fathered at least seven children, only three of whom survived him upon his death in 1661. While his name appears regularly within guild minutes and the standard notarial documents of his day (marriage, business, and apprenticeship contracts), there is little within the civil record to suggest a life outside the ordinary passages of the early modern urban craftsman.

Two qualifications should be made to this last statement. First, Parets was an active participant in local politics. This in itself was not that unusual for a master artisan in Barcelona, as its civic regime, unlike most others in early modern Europe, which excluded craftsmen from any but the most subordinate roles in municipal government, continued to devolve substantial responsibility upon artisans.[26]

Menestrals, or members of lesser guilds like the tanners, not only served on the plenary Council of One Hundred, but also participated in the annual lottery for the six offices of city councillors. Three of these were reserved for the urban elite, while the remaining three were divided equally among merchants, upper guild masters, and the *menestrals*. It is therefore not too surprising to find Parets deeply involved in local politics, especially during the later years of his life. Following the death of his father, Parets was declared eligible for the lottery for public office and served several times on the city council and various of its committees. It was doubtless in recognition of his experience in such matters that the councillors chose him in 1652 as one of the three masters from the lesser guilds to serve on the committee charged with negotiating the surrender of Barcelona to the besieging Spanish army.[27]

Far more extraordinary was Parets's authorship of a chronicle of his native city. The tanner began to keep a record of the principal events in the capital of Catalonia in March 1626. Its two cluttered volumes provide rich testimony to the drama and pageantry of civic life, from its opening page—devoted to King Philip IV's entry into Barcelona—until March 1660, a year before the tanner's death. The work is a curious mixture of genres. Its structure is that of a typical urban chronicle, that is, an impersonal, annalistic record in strict chronological order of significant events in the life of the city. Like most such books, it tends to focus on political and military events, although it also includes detailed accounts of religious celebrations (especially processions), civic fetes, popular riots, and even some information on food prices and other economic matters.

Two characteristics distinguish Parets's account from other chronicles of this sort. First, the events recorded are strictly limited to Parets's own lifetime. This insistence upon contemporary, and to a large extent eyewitness, reporting closely links his manuscript to the genre of daybook, albeit a retrospective one, as suggested by the unbroken hand of the surviving original and the presence of cross-references between its two volumes. Second, while on the whole the tone of the tanner's narrative is factual and impersonal (especially when compared with lively diaries like that of his contemporary Samuel Pepys), at one significant juncture within his account Parets breaks with the cold objectivity of the chronicle tradition to render a highly moving memoir of acute personal crisis. This—his detailed "journal of the plague year" of 1651—constitutes the central document of this book.

If we know little about Parets's own personal history, we know even less of his motives in composing this chronicle. While he often directly addresses a hypothetical reader, his frank criticism of local authorities makes it hard to believe that he contemplated its eventual publication. The complex history of his endeavor is rendered even more confusing by the existence of several different versions of his text. Two major variants of the Parets chronicle survive to the present day: a two-volume manuscript written in his native language, Catalan,[28] and a near-contemporary translation of this text into Spanish.[29] At least two reasons recommend consulting the Catalan as opposed to the Spanish version of his *dietari*. The first is the obvious question of authenticity. Although the Spanish text was published in the 1880s and thus is the rendering most familiar to modern historians, there is no absolute certainty as to the true authorship of the translation. The inclusion of errors involving Parets or persons known to him suggests that the tanner himself could not have been responsible for the Spanish version. The second reason involves the existence of significant differences between the two texts. The anonymous author of the Spanish adaptation of the manuscript did not limit himself to translating Parets's work. On the contrary, in addition to altering the style of the chronicle, he also suppressed the series of brief entries regarding the artisan's personal life found at the end of both volumes of the Catalan version [I, 140v.–146v.; II, 195v.]. These notes recount, in chronological order, significant events affecting both his guild and his family. It is in fact here [I, 141r.–141v.] that one finds the most moving passage in the entire document—the detailed account of the death of his wife and three of their four children during the epidemic of 1651.

Parets's report on January 8, 1651 of the rumor that plague had been discovered in Barcelona was not his first experience with epidemics. During the great scare of 1631–1632, fear of the "Milan contagion" led the local viceroy to admonish Barcelona's inhabitants against those "so pernicious and detestable persons, enemies of human kind who . . . have entered Spain in order to infect the kingdom with poisonous powders as they did in Milan, causing innumerable deaths."[30] As a result, Parets and other members of his guild spent several nights standing guard on the city walls under the orders of the municipal plague board—a tiresome chore, duly noted in his memoirs [I, 144r–v.]. Barcelona was spared visitation by the plague until 1651, when it struck in the neighborhood of Saint Paul's, one of the city's poorest quarters. Although reports of the plague's "entry" into the city began to circulate shortly after the New Year of 1651, local officials hesitated to impose a full quarantine upon the city. As at first only the more impoverished of Barcelona's inhabitants fell ill, physicians attributed the disease to the poor diet caused by the high price of bread [II, 40r. and 41r.]. Such wishful thinking did not last long, however. Even before the full moon of January, which was eagerly watched for portents of an early end to the sickness, many Barcelonans had already fled the city for shelter in the neighboring countryside.

"There was not a single church nor monastery which did not carry out devotions . . . but Our Lord was so angered by our sins that the more devotions were carried out, the more the plague spread" [II, 41v.]. Thus during the weeks that followed, Barcelona's rulers took steps to prevent people from congregating in public. In particular, the city council banned the annual Holy Week processions and pageants, as "this disease does not favor gatherings of people" [II, 41v.].[31] In the meantime, the number of sick in the pesthouse swelled to over 4,000, a figure which did not include those residents whom city officials ordered quarantined in their homes [II, 44r.]. By June, Barcelona had been so emptied of its inhabitants that "it was terrible and piteous to walk through the streets, as one hardly ran into anyone else except someone looking for food for the sick" [II, 44v.]. Shortly after the death of his wife and children, Parets himself fled the city on June 9, accompanied by his mother and lone surviving son. They remained in the nearby countryside at the farm of his brother-in-law until August 4. Then, "Our Lord now worked . . . a most miraculous thing and a great prodigy, that with the plague raging so furiously . . . it was the will of Our Lord to relieve the plague in Barcelona" [II, 48r.]. The city slowly began to return to normalcy despite rumors that "more than thirty thousand persons had died, and some thought that it was as many as thirty-six thousand" [II, 48r.]. After noting the closing of the pesthouse in mid-September [II, 48v.], the tanner mentioned the plague only sporadically in the rest of his chronicle.

One immediately notices clear limits to the uniqueness of the Parets document. Even a hasty glance at other plague journals reveals important similarities, both in content and style, between the tanner's narrative and accounts written by contemporaries from other walks of life. The personalization of responsibility for the introduction and transmission of the epidemic; close identification of plague with the lower depths, both physical and human, of the city; unrelenting emphasis on transgression of the legal and moral order as a crucial agent in the spread of disease—these are but a few of the leitmotifs which appear not only in the Parets memoirs, but also in numerous other early modern plague journals.

Doubtless some of these similarities can be traced to the tanner's own familiarity with the commonplaces reproduced in vernacular medical literature. It is hardly insignificant that the only book Parets explicitly cited in his chronicle (II, 45r.) was the plague treatise published during his youth by the prominent local physician Joan Francesch Rossell.[32] Beyond such direct lines of filiation and influence, it is clear that early modern records of epidemics drew upon a shared vocabulary. Thus, for example, biblical (especially Old Testament) references to plague and other contagious diseases provided generations of fresh witnesses with a common stock of metaphors and narrative devices.[33] For the elite in particular, classical literature also served as an important source of vocabulary and descriptive language regarding the plague. For example, Michel de Montaigne took care to cite Horace in his account of his flight through southwestern France from the Bordeaux contagion of 1585, while the learned Bishop Sprat rushed into print, hot on the heels of the London outbreak of 1665, yet another translation of Thucydides' account of the plague.[34] In addition, many early modern writers betrayed a close familiarity not just with classical narratives of the plague but also with the accounts of writers closer in time, the most widely circulated of which was the *Decameron*. The consequence was a shared stock of images and rhetorical devices—sufficiently shared, in fact, to lead the disease-weary reader to suspect that if you've seen one plague account, you've seen them all.

This would be a hasty conclusion, as there are features of the Parets account which evoke few parallels elsewhere. Some can be attributed to his own idiosyncratic interests and stylistic resources, while others derive from the specific characteristics of the epidemic (and reactions to it) in Barcelona itself. Comparison of his narrative with other plague journals—especially contemporary accounts from mid-seventeenth-century Spain and Italy—suggests three closely related themes as deserving particular emphasis: collective sin, community and outsiders, and class consciousness.

Recent studies in the history of civic responses to disaster have emphasized the importance of collective efforts to restore and reinforce the social cohesion and normal patterns of interchange disrupted by the visitation of catastrophe.[35] This was often achieved by reestablishing spiritual linkages with the divine through highly ritualized acts of penance and propitiation, including (though not limited to) formal vows of commendation to long-standing or newly created patron saints.

That penance should be the appropriate response (both public and private) to such disasters derived from the widespread belief that plagues and other epidemics were acts of divine punishment for earthly sin. In theory, the city was chastised for the misdeeds of all its citizens; hence contemporary emphasis on the collective, even communal, nature of sin and of its remedy.[36] In practice, however, specific groups within urban society were portrayed with growing frequency as especially prone to the violation of human and divine law that allegedly caused the plague. In the context of the Mediterranean epidemics of the mid-seventeenth century, offense against the moral order was usually identified with two groups: outsiders (foreigners as well as local members of ethnic minorities) and the poor, both settled and transient.

Numerous examples illustrate the extensive overlap between these two social categories. From the beginning, Parets, like most other plague writers, attributed the cause of the disease to collective sin: "We had all the greatest plagues in Catalonia . . . thanks to our sins, for a thousand outrages and murders were committed . . . as if it was nothing, having so little fear of God or of the law" [II, 34v.]. Moreover, by locating the initial nucleus of contagion within the poorer quarters of the city [II, 39v.], Parets echoed a theme found elsewhere. In sixteenth-century Seville, blacks were accused of being carriers of disease, while in Padua plague was attributed to the presence of German—that is, Protestant—students. The Cartagena epidemic of 1648 was blamed on an itinerant "African" from nearby Orihuela who sold hemp, and in Rome a decade later the contagion was believed to have been introduced by a soldier and a fishmonger from Trastevere. Not surprisingly, plague was habitually identified with specific areas of cities, especially with the poorer outlying quarters. Hence the association of disease in Seville in 1649 with Triana, the home of the "plebeian sort" who stole "infected clothing"; the close links noted by local physicians between the plague and the impoverished Santa Marina quarter of Córdoba in 1651; and the persistent identification of epidemics with Whitechapel and other eastern parishes in London.[37] Not surprisingly, a frequent result of this linkage between plague and specific social groups was a marked increase in repressive measures directed not only against vagabonds but also against native indigents—measures explicitly justified in plague chronicles.[38]

Thus while most early modern plague chronicles emphasized collective sin as the underlying cause of epidemics, they also invariably charged outsiders—in a social, ethnic, and geographical sense—with actually introducing the contagion. The majority of plague texts also went on to identify criminal behavior as a crucial agent in further communicating the disease, and Parets's chronicle was no exception. He quickly highlighted the moral shortcomings of venal officials as responsible for the plague's diffusion: "In these matters there are always evil persons who cause harm, [and] there were some among those charged with cleaning these [infected] houses who stole some clothes and other possessions. These contaminated other houses, and some persons died as a result" [II, 39v.].

It is precisely at this juncture—the identification of those whose conduct favored the persistence of plague after its introduction—that personal accounts of epidemics begin to vary significantly in content.

This was partly a reflection of the broad range of collective responses to the plague in different cities. For example, at no point in the Parets narrative were the miscreants identified with any particular ethnic group or political faction, which reflects the absence in Barcelona of measures against foreigners, prostitutes, and other offenders against the civic order. There repression was limited to punishment for acts traditionally regarded as criminal, such as breaking and entering, and the sale of stolen goods. The contrast with other plague narratives, especially certain of the contemporary Italian chronicles, is intriguing. Although Milan in 1630, Barcelona in 1651, and Naples in 1656 were all cities threatened by foreign invasion and even immediate siege, only the two Italian cities betrayed a clear predilection for scapegoating by blaming the plague on the enemies of the state, especially the French. The Barcelona of Parets saw neither collective scapegoating nor the association of plague with the machinations of political enemies. In particular, the 1651 epidemic did not give rise to repressive measures against the growing pro-Castilian faction within the war-weary populace.[39] Instead, the tanner (along with his fellow citizens) initially identified a neutral category of "evil persons" (*gent dolenta* [II, 39v.]) as the principal means of transmission of the disease. In this respect, the experience of Barcelona was not that unique. By the mid-seventeenth century, many European cities had come a long way since the pogroms and other acts of collective persecution which had accompanied the Black Death.[40] In fact, the morbid fascination with which the notorious episode of the *untori* and other tales of collective poisoning were received outside of Italy suggests that, if anything, it was Milan and Naples which constituted the exceptions to the increasingly routine character of plague experience during the early modern period.[41]

Yet political neutrality did not entail class neutrality. Here one senses once again the importance of the distinction between the attribution of responsibility for the diffusion as opposed to the introduction of plague. While virtually all plague narratives blame the origins of epidemics on the poor and other marginal inhabitants, the identification of the party guilty of spreading disease frequently expanded to other social groups. The class experience of chroniclers often exercised a strong (albeit not the only) influence on this identification. The point of view of most members of urban elites throughout Europe was predictable: the poor bore responsibility for the triumph of contagion. Ruling class accounts invariably credit not just the origins but also the spread of the plague to the modes of contact among the popular classes. As a result, the "plague dictatorship" erected by local authorities attempted to combat the epidemic by disrupting the standard patterns of plebeian and interclass sociability.[42]

Popular perceptions of disease, on the other hand, often differed substantially from those of the elite. What were crimes according to the ruling class were in the eyes of the lower classes timeworn practices and strategies of survival arising from a broad range of beliefs, interests, and expectations. And in certain instances—for example, during the Milanese and Neapolitan epidemics—these differences in perspective led to overt class conflict. As rulers were all too aware, popular expressions of hatred of foreigners which emerged in these cities during the contagion overlay considerable lower class resentment of the local privileged classes as well. Both government correspondence and plague chronicles from seventeenth-century Italy explicitly acknowledge the popular social and economic resentments underlying accusations of poisoning; some, in fact, express fear of the explosive mixture of popular "superstition" and rebelliousness. Giuseppe Ripamonti's famous narrative of the Milan epidemic of 1630 illustrates elite perspectives on class tensions during times of plague, and other documents report on popular behavior and attitudes in a similar fashion.[43] For example, in 1656 the nuncio Giulio Spinola remarked the widespread belief among the poorer classes in Naples that their Spanish overlords, along with local health administrators and the rich, shared responsibility for the plague. He moreover candidly noted that government officials themselves sought to counteract this dangerous rumor by blaming the French and revolutionary exiles for spreading infected powders. Ironically, what began as a popular rumor in Milan in 1630 was deliberately transformed into an accusation of elite origin in Naples two decades later.[44]

The unmistakable increase in social tensions during times of plague often affords exceptional opportunities to examine the collective comportment of distinct groups; however, it is far more difficult to find adequate sources for the study of popular attitudes and perceptions than for those of the elite.[45] The account reproduced herein by an author of incontestably artisanal origins, provides direct access to the opinions and emotions of the popular classes. It moreover raises questions regarding the general notions of popular culture and class consciousness, questions posed most acutely by the presence of three crucial transformations within the text. Once again, comparison with other plague accounts makes clearer the meaning of these shifts in both content and style.

Like most plague authors, Parets begins by personalizing the disease. On January 8, he learns of reports of deaths "in a house on New Street of a blind man named Martín de Langa" and that "some of the persons who had come into contact with [Langa and his family] also fell ill." A retailer in the local marketplace took sick, he reports, apparently after contracting the disease in a nearby town. "It was also said that several days earlier, plague had been found in Barcelona in a couple of houses but that it had been hushed up. One of the houses was that of the lawyer Tristany, behind Saint Justus. . . . The other was the house of a shoemaker named Matas who lived in an alley at the end of the Passageway of the Prison" [II, 39v.–40r.]. In other words, each individual link in a deadly chain of contacts is personally identified. Thus everyday intercourse among families and neighbors as well as criminal conduct (such as the theft of the goods of the plague-stricken) are identified as the leading agents of transmission of the disease.

Soon, however, the narrative undergoes an alteration. The ravages of the epidemic are shown to be the product not of the continuation but of the *breakdown* of the normal relations of friendship, neighborhood, and family and kin obligations, which had previously united and bound the urban community. The triumph of plague becomes the triumph of egoism.[46] This change of tone begins with a denunciation of the magistrates charged with assuring the city's grain provisions. By accepting bribes and winking at violations of their own plague orders, "these [magistrates] cared only about looking after themselves instead of the welfare of the city" [II, 43v.]. In the following pages—especially in the lengthy section entitled "Of the

Great Sufferings the Plague Caused"—the tale slowly builds to a dramatic crescendo. Parets cites example after example of the sordid victory of private interest over the public good, not just among the lower classes but at all levels of society, including the privileged and the powerful. "The provision of food during the plague was badly organized both in the pesthouse and in the city. Many died in the pesthouse for lack of food who would not have died had things been better organized, and there were many who after arriving in the pesthouse died without having been seen by a doctor or surgeon" [II, 44v.]. The reader follows Parets into a heartrending spiral of crimes, negligence, and the abandonment of the most sacred of social obligations, those within the family:

> I say that it is quite right to flee in order not to suffer from this disease, for it is most cruel, but it is just as right to flee in order not to witness the travails and misfortunes and privations that are suffered wherever the plague is found, which are more than any person can stand. One saw that when anyone fell sick, he lost all touch with friends and relatives, as there was no one who would risk contact with him, just the person nursing him. And it would have to be someone very close or related to him who would dare to take care of him, like a wife to a husband or a mother to a son or a sister to a brother, and even among these many fled or did not want to stay, for the plague was so evil and of such a bad sort that everyone fled. I can give good account of all this, because my wife fell ill with a plague boil on her leg and another on her thigh, from which she died. And although she had two sisters in Barcelona, neither was willing to come to nurse her. . . . [S]he . . . wanted to talk with them and see them before she died, [but] there was no way to convince them to come, as everyone fled from the plague. And there were many cases of this sort. . . . The sick had to find someone, man or woman, who would nurse them for pay . . . all this could be accomplished only by paying money. [II, 45r.–v.]

The author goes on to draw a parallel between the suspension of relations of mutual trust and assistance within the family and the abandonment of obligation within society at large; he sees both as prolonging or deepening the sufferings of the stricken. And, like other plague writers, Parets suggests a notion of duty—what one could call his "plague ethics"—that is firmly rooted in gender.[47] Just as citizens from all social ranks abandoned their particular class responsibilities during the plague, women as well as men failed to discharge their customary duties, as the statement above and the succeeding sections on wet nurses and midwives [45v.–46r.] make evident. The tanner's expectations are firmly rooted in an economy which traditionally charged women with care of the infirm and incapacitated—a domestic role which intensified during times of collective disaster by being inserted into a more extensive cadre of public responsibility. As a result, men are criticized for their failings in institutional roles—that is, for dereliction and abuse of authority—while women are chastised for negligence in the more private setting of family and the household.

Bitter cries of abandonment are, to be sure, a standard feature of plague rhetoric. Yet I would argue that Parets's frequent references to selfish conduct can be read in a slightly different key than the classic terms pioneered by Boccaccio and repeated by generations of plague writers thereafter.[48] A significant clue to finding this key resides in the second transformation of the text: the shift in the social background of those identified as transgressors. In the latter half of his memoir, Parets reverses the viewpoint found in most ruling class accounts. While he continues to charge the urban margin with the predictable crimes of pillage and rape, he also accuses diverse sectors of the local elite with negligence and abuse of authority in pursuit of their own selfish interests. One need only contrast his indictment of members of Barcelona's governing classes for their profiteering and abandonment of duty with, say, Defoe's consistent defense—ironically, in the form of a journal written by a saddler—of the wisdom of the measures decreed by the London authorities, who are depicted in highly charged language as a bulwark of enlightened authority amid a sea of absurd popular superstition. Or, to cite an example closer to home, compare Gavaldá's encomium of the exemplary behavior of the Valencian clergy with Parets's denunciation of the abandonment of the Barcelona sick by their spiritual ministers [II, 46r.]. It is here, in his criticism of the local ruling class, that the tanner parts company with many other plague writers. The greed and carelessness of the urban poor chastised in the first half of his narrative are joined by the corruption and cowardice of the powerful, denounced in the second half. His moral economy is not focused exclusively on the shortcomings of the lower depths, as was the norm in accounts by the elite. Instead, he subjects a broad range of transgressions to scrutiny and condemnation.[49]

Society is hardly an undifferentiated whole in the eyes of Parets. Passages in his narrative suggest that he possessed a firm class identity, based upon a deeply felt separation from both the elite above him and the poor below. This social position was fraught with ambiguity, as it altered according to circumstance; shifts in external conditions often provoked changes in his sense of distance from either pole. This ambivalence found direct expression in his memoir, which should caution the reader against making too stark a distinction between popular and elite plague accounts. As noted earlier, both bodies of writings share many features in common, including their vocabulary, imagery, and sources of reference. Not surprisingly, political perceptions and social attitudes in these two bodies of writing overlap as well, which raises the issue of whether plague texts written by members of the popular classes possess unique, defining characteristics which enable one to isolate the traits of a common popular culture.

One way of answering this question is to examine the conditions in which such texts were produced. First, it is necessary, to recognize the many types of social niches within classes and interactions between classes that helped determine popular culture and creativity. Central to plague (and other) narratives was the use of the written word, that is, the fact that the author was literate and expressed himself through a medium (in this case, a chronicle) habitually dominated by the elite. In the eyes of some scholars, Parets's choice of medium would disqualify his narrative from consideration as an authentically popular statement, regardless of his artisan background. This position—a purism which defines "authentic" popular culture as strictly oral—regards literacy as a skill monopolized exclusively by the ruling class and on occasion depicts it as forming part of a calculated strategy of social control imposed from above.[50] In so doing, it overlooks or minimizes the centrality of literacy as a dimension of the cultural experience of the popular classes in early modern

Europe.[51] It would clearly be misleading to limit artisanal culture, or even popular culture in general, exclusively to the domain of orality. The ability to read and write was not class-neutral, but its possession did not confer automatic membership in the elite. Cultural creation within complex societies like early modern cities habitually took place through exchange, borrowing, and interplay across class boundaries. Not surprisingly, this pattern of extensive and ongoing interchange certainly left its mark on artisan narratives like that of Parets.

Literacy was not the only form of class interaction conditioning popular literary expressions like Parets's chronicle. The possibility of circulating these texts as manuscripts or through the medium of print also influenced the structure and especially the content of these narratives. One measure of this influence can be found in the expression of social and political criticism.[52] One might suppose upon reading Parets's text that popular narratives generally betrayed an impressive willingness to venture criticisms of political and social superiors. We have already seen that the tanner did not hesitate to condemn members of the local ruling class, both lay and clerical, for their corruption and cowardice during the plague. It is in fact not hard to find other examples of the same tendency, such as the late sixteenth-century tailor, Bastiano Arditi, who subjected the Medici rulers of Florence to withering scorn in his private diary.[53] Yet criticism of the dereliction of public authorities can also be found in many elite narratives, for example the harsh words the *regidor* (councilman) Andrés de Cañas aimed at physicians and secular clergy in Burgos during the epidemic of 1599.[54] Meanwhile, some popular narratives contain words of praise for local rulers, as evidenced by the favorable treatment accorded Archbishop Charles Borromeo within both popular narratives of the Milan plague of 1576—one written by the *popolano* Giovan Ambrosio de' Cozzi and the other by the carpenter Giambattista Casale.[55] On balance, an author's intentions regarding the text—that is, whether it was written with eventual circulation or publication in mind, or was to be kept in close confidence—may ultimately have exercised as great an influence on the expression of social and especially political criticism within plague narratives as the class background of the author himself.

Weighing the sociopolitical views expressed within popular plague memoirs requires a more nuanced view of public authority than that suggested by the simple opposition between high and low. It is true that the printer Juan Serrano de Vargas lavished unremitting praise upon all the local and national authorities in his account of the 1649 epidemic in Málaga. Yet at the same time he took care to record significant differences over matters of plague policy among the *corregidor*, the *Audiencia*, and Royal Council in Madrid—respectively, the municipal, regional, and central representatives of the crown.[56] Other plague authors from the popular classes similarly distinguished among the different levels and institutional articulations of the ruling class by contrasting the attitudes and behavior of clerical versus lay authority or the secular against regular clergy. Parets himself was quite outspoken in his criticism of the judges of the *Audiencia* [II, 43v.],[57] the secular clergy who abandoned their parishioners in their hour of need [II, 42v.], and the civic officials in charge of the pesthouse and the city's food supplies [II, 47r.]. However, he has nothing negative to say of the Catalan government officials who abandoned the city and goes out of his way to praise the monks who administered the parishes and second pesthouse during the plague. Deliberate distinctions of this sort are a common feature of popular plague narratives. Their presence suggests the authors' awareness of significant differences and contradictions within governing elites and even of the literary tactics of damning either with faint praise or by juxtaposing strong approval of one authority with silence regarding others.

Another frequently invoked measure of the growing distance between popular and elite cultures in early modern Europe, which is raised with special urgency in plague narratives, is the contrasting approaches displayed by the two poles to general questions of cause and explanation. Early modern elites themselves portrayed the contrast with growing frequency as a frontal clash between "science" and "superstition."[58] This difference in approach exhibited by popular and elite plague writers has become commonplace of modern commentators on plague texts. In an influential essay on popular mentalities during the seventeenth century, Pierre Deyon turned to plague narratives for evidence of the growing gap separating "certain popular forms of religious life and the new needs of scientific thought.

. . . [T]wo conceptions of religious life confronted each other in the seventeenth century, one primitive, tainted with naivete and superstition, the other much more abstract and responsive to the needs of an intellectual elite."[59]

More recent studies have suggested an alternative view to this radical distancing between elite and popular cultures. In a comprehensive survey of reactions to plague in early modern England, Paul Slack has argued that despite evidence of increasing secularization among members of the upper classes, no real divide separated popular from elite understandings of the origins of and proper medical responses to plague.[60] Instead, a "confused eclecticism of traditional views" prevailed in both domains, which lends little support to the suggestion that the popular classes were more prone to providentialism and "superstition" than the ruling class.

Although neither author consulted popular plague narratives in his search for evidence of lower-class attitudes toward the etiology and spread of epidemics, such documents shed light upon popular interpretations of plague in specific historical circumstances. As we have seen, Parets repeatedly expressed belief in the divine origins of the plague, a posture not at all in conflict with his attribution of responsibility for the introduction and further transmission of plague to specific individuals and social groups. While he provides relatively little information about the symptoms and clinical evolution of the disease, he does endorse [II, 41v.] the widespread belief in the close relation between the movements of heavenly bodies and the intensity of plague. The same view is also found in another popular record of the same plague, the 1650–1651 diary of the peasant Joan Guàrdia.[61] Much more significant, however, is the fact that the astrological leanings of these texts were not beliefs limited exclusively to the domain of popular culture. Numerous elite commentators on the same epidemic expressed identical views regarding the celestial origins of and influences upon the plague.[62] Contemporary physicians like Gaspar Caldera de Heredia[63] and Alonso de Burgos,[64] writing on plague in mid–seventeenth-century Seville and Córdoba respectively, strongly defended the astral sources of the plague, a stance which fit in nicely not only with mainstream Galenist views on the origins of epidemic disease but also with the position of less traditional works, such as Oliva de Sabuco's *New Philosophy*, long regarded as one of the more innovative treatises in early modern Spanish medicine and natural philosophy.[65] Religious authorities also endorsed this view. For example, the head of the Jesuit college in Huesca, in a report to his superiors in Rome, attributed the origins of plague there in part to the "special constellation caused by the lack of change in the phases of the moon."[66] Finally, the private writings and journals of members of the privileged classes often betrayed firm belief in the influence of the celestial bodies on the evolution of plague. Attribution of astral (and other) origins of the disease are found in the impassioned writings of the French magistrate Pierre Robert.[67] Writing in the early 1630s, Robert linked the appearance of plague in his provincial town to the conjunctions of stars and eclipses, complemented by the workings of the devil and the abundant sins committed by the local canons whom he so thoroughly detested.

Clearly a crucial part of the problem of identifying class attitudes resides in the notion of popular culture itself, a concept cast in terms too broad and undifferentiated not to engender a sense of unease. If anything, the case of Parets's narrative suggests the inadequacy of the simple distinction between "elite" and "popular" as categories for analyzing the complex society of early modern European cities and the multitude of cultural spheres they sheltered. Its reader confronts a document by an author of undeniably popular origins; there can be no doubt of Parets's juridical status as a commoner nor of his membership in the *poble*, the social category which embraced guildsmen and their families. Nor can we question the tanner's ongoing exercise of his trade, one which his fellow citizens, moreover, ranked among the baser of the mechanical arts thanks to its practitioners' unavoidable contact with blood, urine, and other "polluting" materials.[68] His text nevertheless appears to share remarkably few features that could be isolated as exclusive to non-elite plague narratives. In fact, when one examines the larger corpus of popular accounts of epidemics, few truly distinguishing characteristics (apart from specific linguistic variants) can be singled out.[69]

That Parets's and other artisans' accounts resemble elite memoirs as well as other diaries and chronicles of non-elite origin should come as no surprise given the middling position of craftsmen within early modern urban societies. Yet it is precisely the existence of mixed and intermediate roles in society and culture that the opposition of elite versus popular tends to ignore. If the latter term is used to refer to a category including anyone or thing not pertaining to the elite, it invariably becomes too broad and imprecise to be of much use. This does not deny the existence of a deeply rooted collective consciousness on the part of rulers and ruled in early modern cities; on the contrary, inhabitants of early modern urban societies were acutely aware of the boundaries and distinctions separating one rank from another. Rather, I wish merely to suggest that the polarity between high and low, which plays such a prominent role in numerous early modern discourses, tends to ignore or mask the existence of various intermediate levels and categories, of social and cultural hybrids who looked both above and below for their validation, resources, and signs of identity. Many artisans, like Parets, lived in both worlds: between their seats on the city council and their sporadic participation in (or approval of) street politics; between literacy and unforced access to the vivid oral culture of the peasantry and working class (including journeymen and apprentices); and between limited forms of social, economic, and political sharing with the privileged, and a certain (also limited) solidarity with their less fortunate neighbors. Any historical analysis which does not keep these ambiguities and ambivalences in mind risks losing sight of an intriguing range of cultural experience, one which gave forth products of considerable interest and often of genuine importance.

The third and final transformation within the text—and the most striking aspect of the tanner's own response to the disaster—was the dramatic alteration of the character and tone of the document he willed to posterity; the passages on the epidemic of 1651 constitute the only juncture in the two volumes of his narrative in which he abandoned the cool objectivity of the chronicle for the highly personal tone of a memoir. In the face of collective and personal catastrophe, Parets dropped his long-standing habit of borrowing phrases (and on occasion lengthy passages) from pamphlets and other contemporary publications to write a direct, eyewitness account of family tragedy and, more generally, the disintegration of society as he knew it. The assumption of a strong, plaintive voice of his own reveals the extent to which the plague overturned the calm certainties of daily relations, working a transfiguration in the nature of the text itself. As the anonymous editor of the Spanish translation of Parets's account explained, the plague was "the most painful and piteous event which you will find, not only in the two volumes of this chronicle, but in whatever you may read among lamentable and moving tragedies."[70]

It is one of the paradoxes of plague literature that the most detailed and realistic descriptions of the epidemic have been written by authors who lacked personal knowledge of the disease. Such, at least, was the case of Manzoni and Camus, and perhaps even of Defoe and Boccaccio as well. It is equally ironic that, with few exceptions, the accounts written by direct witnesses—that is, survivors—of the plague should prove so cursory and devoid of emotion. One cannot, to be sure, compare the literary strategies and resources of early modern journals with those of the modern novel. Nevertheless, the general absence of sustained personal commentary and emotional response in early modern plague testimonies is striking.[71] Indeed, it baffles modern sensibilities, which expect autobiographies and diaries to provide the most revealing expressions of intimate reflections and sentiments. During the early modern era, however, personal memoirs and daybooks (apart from spiritual autobiographies and records of private meditations[72]) tended to adopt a factual and impersonal mode of reporting. Overtly emotive prose, on the other hand, was more frequently associated with other literary genres: with sermons, for example, or with sentimental typologies such as pastoral romances. The latter in particular could scarcely serve as a likely model for such dire straits as the plague.

In short, a sense of the relative appropriateness of certain rhetorics dominated most of the personal literature written during this period. It has already been noted that Miquel Parets was a minority voice, in that his is one of the few surviving popular accounts of plague. It is therefore not too surprising to find that his testimony departs from the norm in yet another way. It is impregnated with a genuine emotionalism far removed from the stoic reserve characterizing most such eyewitness records. His literally naive account could not be more distant from the impersonality and stilted formality of most other early modern plague narratives, many of which were composed for publication. In the end, the memoirs of this singular tanner presented—like the journal of Dr. Rieux, the quiet but determined hero of Camus's *The Plague*—a straightforward message: "to state quite simply what we learn in a time of pestilence." It is fitting that this introduction should close in Parets's own words. Toward the end of the summer of 1651, as the disease began to decline in intensity, the tanner offered this final, weary summation [II, 47r.] of the "sad spectacle"[73] of the plague year:

> Of all this I can give good account, after having gone through so much suffering from the plague within Barcelona, with my wife dying and a son almost thirteen years old and another who was eleven and a little girl one year old whom my wife was still nursing, four persons in all, all dying in less than a month. My mother and a son four years old also caught the plague and recovered, and all these dead and sick persons were cared for in my house and I can give good account of what these sicknesses cost. . . . I witnessed the many travails of many persons . . . so many that it would take a long time to tell it, but you should be able to imagine it thanks to what little I have already written. . . .

A Note on the Translation

That present-day Catalans find it hard to understand many of the passages in Miquel Parets's chronicle—so much so, in fact, that the recent edition of the plague narrative of 1651[1] had to be substantially modified for the modern reading public—should give some idea of the difficulty of translating this text. Given the impossibility of conveying to the English reader many of the peculiar qualities of the tanner's prose, I felt it advisable to try to identify certain of its more significant features. While some of these can be found in other Catalan documents from the same period, others are of Parets's own devising.

1. Although the chronicle contains relatively little repetition in matters of content, the tanner constantly reproduces the same turns of phrase. In fact, it is not uncommon for him to repeat the same word or phrase once or twice within the same sentence. For example, by far his favorite adjective is *gran* ("great" or "large," although it could also mean "old"). I have taken the liberty of pressing into service a wide range of modifiers to render this simple if tedious word into English.
2. Parets's grammar is relatively uncomplicated. However, it is often difficult to distinguish his subjects; sometimes they disappear altogether, thanks to his (rather frequent) use either of an unidentified "they" or the passive voice. I have retained the latter feature, while usually substituting a more precise noun for the undifferentiated pronoun. The craftsman also frequently confuses the present participle with the conjugated verb, a trait I have not seen in many other contemporary documents.
3. Most of Parets's sentences begin with "and," a stylistic usage hardly unique to his text; for example, the autobiography and other writings of the London astrologer Simon Forman share the same

tendency.[2] Also, his syntax was often excessively complicated, usually the result of his penchant for long, run-on sentences. These in turn derived from his cheerful disinterest in punctuation and his aversion to paragraphs (again, characteristics found in many early modern texts).

4. Parets's spelling is highly irregular, especially when rendering proper names, which suggests a certain oral, even phonetic, quality in his writing.

5. Consistency does not shine among Parets's virtues as a writer. For example, at one point [II, 40v.] he states that after some thieves were hanged "no one spoke of committing any crimes" in Barcelona. He then devotes the next section to recounting a murder committed shortly thereafter. It is clear in this instance that the tanner's approval of the tough law-and-order measures the city government carried out led him to write a sentence contradicting the evidence he himself provides in the rest of the chronicle regarding continued law-breaking during the plague.

6. As in most seventeenth-century Catalan texts, there are substantial borrowings from contemporary Spanish ("barbarisms," to use the loaded designation of modern Catalan linguists). I have the impression, however, that Parets's Castilianisms were not too extensive, especially in comparison with elite writings from the same period.[3]

Like most translators, I have attempted to reconcile the conflicting demands of readability and accuracy. While I have done my best to provide as faithful a rendition as possible, I have nevertheless devoted most of my efforts to making the text comprehensible to the present-day reader. It almost goes without saying that I am now more acutely aware of the difficulties of translation than when I began this project. I am therefore particularly grateful to Xavier Torres, my co-editor in the preparation of the Catalan edition of this text, and Xavier Gil for their help in interpreting many of its more stubborn passages. I should also mention that my other authority in matters linguistic is the etymological dictionary edited by Father Antoni M. Alcover et al., an indispensable source for medieval and early modern Catalan usages.[4] Despite extensive reliance on all three sources, any remaining errors are my exclusive responsibility.

MIQUEL PARETS

A Journal of the Plague Year

Plague in Catalonia in the Year 1650

By European standards, Barcelona in the mid-seventeenth century was a middling-sized city, housing some forty to fifty thousand inhabitants.[1] *Its economy rested on a broad base, which included commerce (both local and maritime), agriculture, and a wide variety of crafts and small-scale industries. While not as prosperous a city as it had been during the late Middle Ages, when Barcelona rivaled Genoa for control of long-distance trade in the western Mediterranean, it nevertheless continued to dominate a dynamic regional economy. Moreover, as capital of the Principality of Catalonia, one of the many territories comprising the Spanish monarchy, Barcelona played an unusually active role in peninsular and even international politics. And politics was very much on the minds of its citizens in 1651, when the plague struck.*

Relations between the citizens of Catalonia and the court in Madrid grew increasingly tense during the 1620s and 1630s, as the monarchy's persistent demands for taxes and troops for the war effort against France met growing resistance from Catalans of all classes. On Corpus Christi Day, 1640, popular revolt within Barcelona (which included the assassination of the viceroy, the highest ranking royal official in the Principality) brought matters to the breaking point. Shortly thereafter, war broke out between the Catalans and the crown. After an ill-fated attempt to declare independence, the Catalan government placed the Principality under the protection of France in 1641 and recognized Louis XIII as its sovereign.

Rebellion and war—these were the grim realities of the decade spanning from the uprising known as "Bloody Corpus Day" to the appearance of yet another civic disaster, plague. The Barcelona epidemic of 1651 began as an extension of the Valencian "contagion" (to use the contemporary term) of the later 1640s. It thus entered Catalonia from the south and was most probably carried by troops who returned to the city of Tortosa[2] *after having sacked villages in the northern reaches of the Kingdom of Valencia. Miquel Parets's earliest comment on the plague dates from 1650. Significantly, he marked this passage with the first margin note—"Plague in Catalonia in the year 1650"—to appear in his chronicle, which suggests the special importance he attached to it.*

[II, 30v.] Following the enemy's retreat, the commander of the Catalan cavalry, Don Joseph d'Ardena,[3] was sent along with a thousand horsemen and a thousand foot soldiers to the Kingdom of Valencia, where they sacked the towns of Peñíscola, Benicarló, and Sant Mateu.[4] While they did much damage there, it also turned out badly for Catalonia (as will be seen below), because thanks to this raid they brought the plague to Tortosa, the Ebro River valley, Tarragona,[5] and other places. That summer it had afflicted the entire Kingdom of Valencia and it was not known whether or not it had died out. The horsemen sacked some villages there whose inhabitants had fled from the disease, and since soldiers always rob without thinking of the harmful consequences that follow, they brought much booty still infested with plague to Tortosa. Tortosa was thus the first town it struck, as it was closest to Valencia.

The plague quickly extended throughout the Ebro River valley and soon reached Tarragona, the largest city in southern Catalonia. In the meantime, Barcelona's authorities began to take protective measures against its spread.

❧ [33r.] Of the Measures the City of Barcelona Took Against the Contagion in Tortosa

It was noted above that the expedition of Don Joseph d'Ardena and the Catalan cavalry to sack the Kingdom of Valencia brought the plague to Tortosa.[6] And that the year before, in 1649, pestilence had prevailed throughout that kingdom. And that since this disease tends by its very nature to lie dormant, just when one thinks that it has departed it flames up again.[7] . . . [Thus when the soldiers] entered into villages whose inhabitants had not returned out of fear that the plague was still alive, and since soldiers pay no attention when they are robbing, they did not hesitate to loot these villages and bring back to Tortosa everything they could, including those things which still carried the plague.[8] Thus once they got back to Tortosa the illness awakened there, and there was no choice but to declare that plague was in the city.

Seeing this, the city of Barcelona quickly banned trade with Tortosa and other towns in the Ebro River valley, for the plague had already spread to the nearby villages.[9] It also sent a physician and a barber-surgeon to see if the disease was plague. The physician was Dr. March Gelpí and the surgeon was Master Mates of Broad Street, both of whom were well paid for their trip. And when returning they took a boat up the Ebro in order to avoid a bad mountain road, and were captured at a narrow pass by a patrol of enemy guerrillas and taken as prisoners into Aragon. They [the guerrillas] asked a handsome ransom for them, 1,500 ducats (at that time a ducat was worth 8 pounds 10 shillings in copper coinage).[10] In order to negotiate it they freed the barber Master Mates and kept the doctor. They fixed such a high price because they knew that the doctor and the surgeon were emissaries of Barcelona and that the city would have to pay their ransom. The viceroy [of Barcelona] tried to intervene by writing to the viceroy of Valencia and Aragon, arguing that they [the captives] were not prisoners of war but rather that since they were travelling for the welfare of both sides no ransom should be demanded. He also offered to trade some other prisoners, but to no avail, for they [the guerrillas] continued to demand the ransom. The city had no choice but to pay and thus turned over 675 gold ducats, although there was great reluctance to do so, as it [the city] had not authorized them to travel by boat. But in the end the city paid. It also sent another physician, Dr. Vileta, who returned saying that it was definitely plague. The city government set up a watch and refused entry to anyone from the Ebro valley, with or without plague passes.[11] And only four city gates remained open: the New Gate, St. Anthony's, and the Sea and Angel gates.

Plague in Tarragona

At the same time, Lent of 1650, the plague also started up in Tarragona, where a great many persons died. They say that it came in a ship bringing cloths from Valencia, and it became so powerful that a great many persons died. In Vilaseca[12] it spread so much that everyone had to leave the village and take shelter in huts, where people from neighboring villages brought them food. They had to do the same thing in Tarragona; the soldiers took refuge in huts near the sea, and hospitals were set up inside and outside the city. Its [Tarragona's] inhabitants were made to give food to the sick and a great many people died.

The city of Barcelona ordered its monasteries and convents to say prayers for protection against the plague and to display the Holy Sacrament in all the churches, beginning with the Cathedral and then the monasteries and parish churches one by one.[13] Processions were held and prayers were said to placate the anger of Our Lord God.

The plague guards kept a close watch and refused entry to anyone from the countryside around Tarragona, despite the fact that care had been taken not to let anyone leave Tarragona under pain of death, and thus no one had dared leave that city. Barcelona closed all its gates but two, the Sea and Angel gates, through which everyone had to enter and exit. But as the Angel Gate is so narrow and so many carts and mules and people had to enter, there was little room for the guards, and it became so crowded that it was easier to break the rules. Thus after a few days the other two gates were opened again, and they [the guards] had to go back to the same watch as before.

Soon a new menace appeared—a second focus of contagion, this time descending from northern Catalonia and the border with France. Barcelona intensified its preparations, both spiritual and material, as it anxiously awaited the plague's approach.

❧ [34r.] Of the Plague in Girona and Other Parts of the Empordà

In early June 1650 the rumor was heard in Barcelona that the town of Sant Pere Pescador[14] in the Empordà[15] had plague and that everyone had fled, leaving the town abandoned. Plague was spreading throughout the Empordà, and around Corpus Christi Day the news came that it was in Girona[16] and many persons had fallen ill there. Thus the city of Barcelona reinforced its watch, closing tight all its entries except the Sea and Angel gates, and sent a physician named Dr. Vileta to Girona to see whether the disease was plague. He went there and conferred with the local doctors, some of whom said it was plague while others said that it was not. Either because of their unwillingness to declare it such or because they didn't know what it looked like or because of some bribes that people said were paid, Dr. Vileta declared that it was not plague but rather some other disease.[17] He then returned to Barcelona and had the trade embargo against Girona lifted.

It was a great miracle of our Lord that the plague did not enter Barcelona at that time, because since most people in Girona were sure that it was the plague, many took flight and fled to Barcelona. And they were right, for the embargo was lifted for only a week. Afterwards it was declared again, and no one was allowed to enter the city, as things were going badly in Girona, where the disease acted very rapidly and many persons were dying. Thus the city of Barcelona and its Plague Board[18] decided[19] once again to send a physician and surgeon to examine the disease and end the confusion over what it was. They named Dr. Argila and the surgeon Jaume Texidor, considered the two most capable in medical matters in Barcelona.[20] They went to Girona toward the end of July and spent eight days there and decided that it was clearly plague and that there should be no joking about it[21] as it was quite virulent. When they returned they were made to undergo quarantine in a country house near Sarrià belonging to Dr. Argila.[22]

[34v.] The area around Girona was also closely watched so that the disease would not spread to the countryside. People were given rations because the food situation there was critical, as for some time no one would go there or sell them anything. The Royal Council[23] sent the judge Mr. Camps y Rubí,[24] who placed many guards around Girona, along with a line of stakes set a short distance from the city to serve as markers.[25] They [the guards] did not let anyone from Girona pass beyond this line, as they had orders to kill any citizen found outside the boundary. Mr. Camps i Gurí also paid for rations and foodstuffs to be brought in from the nearby villages, and the city government and Cathedral chapter joined together to pay him back. Thus the rations were brought up to one of the stakes, and the money to pay for them was either passed through a fire or washed with vinegar.[26] The citizens asked for what they needed and the goods were brought to them there, and in this fashion the city was fed without anyone having to leave it. The plague raged with great intensity, and there were days during which 150 to 200 persons died. It was mostly their own fault because they had not taken measures when it began.[27] Even when some of the doctors there knew back in April that there was plague in the city, they tried to keep the news from spreading, and by wanting to keep it secret it spread the way it did.

Many villages in the Empordà were quarantined, and there was also plague in Olot,[28] but as they saw from what happened in Girona that it was better to admit it [to having plague] right away, they were given help in cutting off contact with the outside. As a result, the plague did not spread so much. In Barcelona the parish churches and monasteries held many processions, each displaying the Holy Sacrament in turn. Prayers were said not only for the disease. Rather, we had in Catalonia then the three greatest plagues God can inflict upon a people—famine, disease, and war—all of which we had in abundance.[29] Famine struck Barcelona in August, when bread began to get scarce and people crowded at the bakeries looking for it. There are no words to express how the poor suffered, and all this came from people hoarding wheat and not wanting to sell it because the harvest had been so poor.[30] Even in August when wheat was selling for 8 or 9 pounds per *quartera*[31] they still hoarded it, hoping to sell it for much more. Thus the city government had to help distribute bread, as the bakers did not have enough, and the dearth was so great that everyone feared that there would be a riot in Barcelona. But things improved after the city took the dough from the bakers and baked bread at the Customs House,[32] and everything calmed down. And all this happened not only in Barcelona, but also in many places, and we can truly say that we had all the greatest plagues in Catalonia thanks to our sins, for a thousand outrages and murders were committed, so many that they could not be counted. Especially grave were the murders of clergymen, for people did not hesitate to kill priests as if it was nothing, having so little fear of God or of the law. Thus it was not surprising that Our Lord sent so many misfortunes to our country.[33] May He have mercy upon us and be willing to hear the prayers of some righteous person, for among so many evil men there always has to be someone good.

The inevitable did not take long to happen: plague was reported in Barcelona itself on January 8, 1651.

1651
❧ [39v.] When They First Realized the Plague was in Barcelona

On Sunday January 8 of that year,[34] in the house on New Street[35] of a blind man named Martín de Langa,[36] there died a young girl, a relation of his wife, who also quickly fell sick and died. And some of the persons who had come into contact with them also fell ill. When news of this reached the councillors,[37] they went immediately to the house at nighttime in order not to spread the disease or to cause panic. They ordered the sick persons to be removed to the convent of *angels vells*.[38] Everyone who had entered the houses of the sick who had not fallen ill was sent to the towers of Saint Paul and Saint Severus[39] in order to isolate them from contact with others. They also ordered the houses on New Street to be cleaned and fumigated[40] and much clothing there to be burnt.[41] And as in these matters there are always evil persons who cause harm,[42] there were some among those charged with cleaning these houses who stole some clothes and other possessions. These contaminated other houses,[43] and some persons died as a result. At the same time a man from Barcelona named Campderrós, a retailer who lived in the Born marketplace[44] and who had been imprisoned for debt, was brought to the general hospital.[45] He had gone to Olot when the plague was raging there and was sent to the hospital [in Barcelona] when he fell ill with a plague boil. After the doctors and surgeons in the hospital examined him they didn't want to admit him. Rather, they had him taken to the *angels vells* with the others, amid murmurs that an example should be made of him. Later he recovered, and afterwards he tended the stricken[46] who were transferred to the new pesthouse at Jesus Monastery, where he fell sick again and died. Only God knows if it was because he had dared enter Barcelona with such symptoms after coming from a plague-stricken area.

It was also said that several days earlier, [40r.] plague had been found in Barcelona in a couple of houses, but that it had been hushed up. One of the houses was that of the lawyer Tristany,[47] behind Saint Justus,[48] where someone sick with the plague had entered and right away they smuggled him out of the city. The other was the house of a shoemaker named Matas who lived in an alley at the end of the Passageway of the Prison[49] giving on to another alley which joins the Knifemakers' Street, where someone died (they say of this sickness) before Christmas. Clothing also left this house and contaminated some others.

At this time when the sickness was beginning in Barcelona and people were wondering whether it was plague, many persons left the city. There were many among them who had fled from the plague in Girona, and as they were already familiar with the travails it caused, many of them left. Just mentioning the plague of Girona scared others into fleeing, and many took their belongings and clothes to monasteries. Things just went along like this, with one day one person getting sick and another day yet another, and somebody was always being taken to the pesthouse. Most of them were poor people, so it was believed that their misery and the poor diet caused by the famine had made them fall sick.[50] The poor people could not find bread to eat as it was so scarce, not because the city did not bake it, but because everyone from two or three leagues around Barcelona came in search of bread because there was no wheat in the nearby towns and villages. It did no good to put extra guards at the gates to keep people from taking bread out, as people tossed it down from the walls. The poor suffered so much that many days there was no bread to eat, but instead carrots and vegetables and other bad food which cause diseases.[51] Thus when they fell sick their diet was blamed. Meanwhile, many persons were waiting for the full moon to see what would happen, as it always changes position somewhat, and it was full on January 24, although it did not do much.[52] Meanwhile, the disease continued to spread, and people kept fleeing, because those who are the first to go find it easy to find refuge anywhere. For this reason people continued to leave the city.

Of the Soldiers the City Hired to Guard the Walls

When the city saw that so many were leaving and that commerce and industry were doing so badly and that the poor people were suffering so much, the Council of One Hundred[53] decided to raise 700 soldiers from among local guild members[54] to guard the city day and night. Each soldier was paid three *reals* daily,[55] while corporals were to receive four, sergeants five, and second lieutenants six. Captains were not mustered; meanwhile, the soldiers were enlisted and divided in four regiments of 175 each. One regiment was assigned to the east bastion, another to the southern one, another to the bastion of Saint Madrona Wall, and another to the fort at Saint Anthony's Gate, where a large wooden barracks was built. All were led by the same field commanders, sergeant-majors, and adjutants who had previously commanded the city militia, and each regiment was to be joined by two militia companies and their dozen officers. The guard was posted around the clock for two days at a time. During the day they watched the gates and stood reserve for the walls while forty men went up to Montjuich.[56] Only the four major gates were open, and they [the guards] were careful not to let outsiders take out bread, which they confiscated. At night they stood guard on the wall itself, in each sector. The soldiers were signed up without delay, because everyone knew that trouble was coming. Many found employment this way, which was the city's aim, as in this fashion not everyone would flee, and many poor persons who would have fled stayed instead and were taken care of. And this way the city was also very well guarded, thanks to the diligence of the corporals.

❧ Of a Case of Justice Carried Out in Barcelona in Saint James's Square[57]

Law and order were lacking in the city of Barcelona because there was no viceroy.[58] Since no one feared the night watch, many evils and robberies were committed. One couldn't leave one's house at night thanks to the muggings and thefts and murders and such shamelessness that many nights the watch itself was shot at. The apprentices didn't dare [40v.] patrol at night except in well-armed groups of eighteen or twenty, and even then they were afraid, because the thieves were usually soldiers from the local batallion,[59] and thus horrible people without fear of God or the law. And many others, while not soldiers, also dressed themselves in uniforms and robbed and mugged. Barcelona had reached such depths of misfortune that no decent man could leave his house at night without something terrible happening to him. Hence the following incident, well fitting such times.

On January 17, Saint Anthony's Eve, three scoundrels (who weren't soldiers) went to Saint James's Square at seven o'clock in the evening to rob an officer returning from a gaming house. He was accompanied by a Neapolitan named Nardo, who worked as a gambler and had married in Barcelona and lived here many years. And when the Neapolitan saw that they wanted to rob his friend, who was carrying some money in his pocket, he began to cry out, shouting what villainy is this and if they meant to rob them. When they [the thieves] saw that he began to yell, one of the three thieves, a metalworker named Salvador Cotures, stabbed the Neapolitan several times, from which he died on the spot. He also tried to stab the officer, but in the dark he hit one of his fellow thieves, also a metalworker, so badly that he died two or three days later. The officer fled in one direction and the thieves in another, although it was said that three other friends of the robbers were waiting for them in the entry to Saint James's Church.[60] These three had recently been drinking with the thieves in a tavern and were waiting for them there, not knowing of their evil plans. But after they heard the noise they wanted to see what it was. They recognized their friends and then said good-bye, and each one went his own way.

When the officers of the law saw such shamelessness they began to search for the scoundrels, and after finding the metalworker who was wounded they ordered a court scribe to take his deposition under the strictest secrecy. Thus they found out everything and sent out patrols, and on January 19, Saint Sebastian's Eve, they dragged Salvador Cotures from sanctuary in the church of Saint Sebastian.[61] They also arrested his comrade, a glassmaker named Ramon Font, and after imprisoning them they speeded up the trial because the officer made such a fuss about Nardo's death. Thus on Saint Paul's Day, which was January 25, 1651, a criminal trial was held and they were condemned to hang in Saint James's Square at the same spot where they committed their crime, even though there was some protest over Salvador Cotures having been removed from sanctuary.[62] Knowing this, the officers of the law did not wish to allow them enough time to file an appeal. Instead, the thieves were hung by heavy chains in Saint James's Square between eight and nine o'clock that very night. Many people came to see it, and since so many people came one of the city's militia companies was ordered out to make room. Thus they were hanged eight days after committing their crime, and at almost the same time and in the same place. They were left hanging all night, until nine or ten o'clock the next morning, when they were buried. Of the three others mentioned above, two were caught and were not punished at all, because (it was said) the two who were hanged had not implicated them in anything. But the one who informed on the others died two days after being wounded. This rigorous example frightened other thieves, and they left the city, and afterwards no one spoke of committing any crimes.

☙ Another Case of Justice in the Hatters' Street

Shortly after this case another example was made in Barcelona. One night a man treacherously went to the Hatters' Street to the house of a hatter named Poll, which was in front of the steps of Saint Mary's

Church[63] on the corner of the Mirrormakers' Street. He called for the master and suddenly fired a pistol shot and killed him. All this had been prearranged, for he had been paid a hundred *reals* to kill him. Seeing such shamelessness [41r.], and that no one in Barcelona was safe in his own house, a search for this man was ordered, with a reward promised to whomever helped capture him. Thanks to this he was found and brought to trial. And it turned out that the man who had ordered the killing was a fellow hatter named Tiana, who several days before had quarrelled with Poll and threatened him, and had now fled from Barcelona. The murderer, a gardener from the Saint Paul's quarter,[64] was found guilty and condemned to hang. His hand was chopped off in front of Poll's house, and then he was whipped through the Moneychangers' Street and hanged on the sea gallows. This sentence was carried out in the month of February 1651 and served as a lesson for others.

❧ How They Carried the Body of Saint Madrona in Processions to Implore Rain

There was great famine in Barcelona and most of Catalonia, so much so that no wheat could be found in the countryside. It was so expensive that at the beginning of Lent wheat sold for 10 and 12 pounds a *quartera* and right before that year's harvest it reached more than 20 pounds a *quartera*, and even then it was hard to obtain even from good friends. Most people in Barcelona and from the surrounding area bought bread within the city, even though there was such a crowd that were it to revolt it would have been impossible for the city to provide enough bread for everyone to eat. So much bread was smuggled out of the city by land and sea that guards had to be placed at the gates to search those who left and to make them leave their bread for the city's inhabitants, who were suffering, and the city could not supply everyone. And the bread was made in tiny loaves, and it was forbidden to make them any larger for they were all 4 pence would buy. And with all the dearth of wheat, none was found at this time because the harvest was in poor shape thanks to the drought. For that reason they resorted to the usual solution of carrying the body of glorious Saint Madrona to the Cathedral.[65] They went to fetch it on

March 5, 1651, and as it was so close to her feast day it seemed improper that she should be away from her own home on that day. Thus the councillors promised right there that if the glorious saint should provide us with water in abundance she would be returned home on her feast day. The glorious saint heard the prayer of some good soul because the day after she went to the Cathedral Our Lord was pleased to give us abundant rain one or two times. And while for some reason she could not be returned on her feast day, the following day she was returned with a *Te Deum Laudamus* along with great joy on the part of everyone. Truly the weather and the harvest were doing poorly, and thanks to the rain, things improved greatly. Praise the Lord, through glorious Saint Madrona.

How the Plague Continued in Barcelona

You will find on the next to last page how the contagion started in Barcelona, and how steps were taken to deal with it, and how in order not to scare people they did not dare pronounce it plague. Instead, they covered it up by saying that those who died were all poor people and that it all came from the bad food they ate last winter. Thanks to the famine they could not find bread, and what they suffered was a great pity, as often they had to go to bed without having eaten bread all day. Since they had to fill their stomachs with one thing or another, many ate nothing but carrots and cabbage and leaves, which was bad food and made many sick. Thus many persons began to die in one part or another of the city, and in order not to frighten people they removed those suspected of having the plague to the convent of the *angels vells*, which is a large building next to the garbage ditch outside Saint Daniel's Gate. There they had plague doctors and surgeons and monks to administer the sacraments and wards and everything else they needed. They took them there at night under the supervision of the plague warden, who was Master Coll, the broker.[66] People became very frightened and many fled Barcelona, [41v.] leaving their better belongings in monasteries or in other hiding places. And since this disease normally is affected by the phases of the moon, many waited to watch the full moon of February. It changed position slightly, but to no great extent.

After this a great many people left Barcelona, and as the number of sick in the city kept growing and the aforementioned pesthouse of *angels vells* was small and (it was also said) unhealthy because of the nearby swamps, they decided to move the pesthouse to the Franciscan monastery of Jesus.[67] Thus in early February they moved all the sick from the *angels*. Those who could, walked, and those who could not rode in carts, and at that time there was room enough for all of them. The friars left the monastery, some fleeing quite far, but most of them went to the farm of Remanyà, a peasant from Sarrià,[68] located right above the monastery. The city provided a great many beds and straw mattresses as the number of sick grew, but in spite of all this they still didn't want to admit the presence of plague instead of other diseases. But when the full moon came, many persons realized what was happening, because so many were dying and many fell sick with different sorts of plague boils, which looked like beans, all red with a black tip.[69] After the full moon, a great many persons left Barcelona, and they took with them much of their clothing. They left their best things with monasteries or hid them in their own houses in order not to mix them with the clothes they used daily, just in case plague was round in their houses. And among these waves of refugees two thirds of Barcelona's inhabitants fled, which stunned those who stayed there.

There are no words to describe the prayers and processions carried out in Barcelona, and the crowds of penitents and young girls with crosses who marched through the city saying their devotions.[70] The streets were constantly full of people, many greatly devout and carrying candles and crying out "Lord God, have mercy!" It would have softened the heart of anyone to see so many people gathered together and so many little girls, all of them barefoot. To see so many processions of clergy and monks and nuns carrying so many crosses and so many rogations that there was not a single church nor monastery which did not carry out processions both inside and outside their buildings. But Our Lord was so angered by our sins that the more processions were carried out the more the plague spread. And this year they didn't hold processions on either Maundy Thursday or Good Friday in order to avoid any crowding of people,[71] and on the eve of Maundy Thursday they closed the churches at ten o'clock and ordered them not to open until the next morning. They also prohibited worship at the Easter altars[72] so that people would not come into close contact with each other, because this disease does not favor gatherings of people.

How the City of Barcelona Chose Saint Francis de Paul[73] as Its Protector Against the Plague

Great devotions and vows were made in the city of Barcelona by individual citizens as well as by the municipal government.[74] Thus the city of Barcelona, at the suggestion of some good persons, decided during a meeting of the Council of One Hundred on March 26, 1651, to make a vow to take glorious Saint Francis de Paul as its protector against the plague. And that every year on his feast day, which is April 2, a celebration would be held and that the Cathedral chapter would form a procession to his church to say mass there just as is done on Saint Raymond's Day.[75] And to this end, on Monday, March 27 of that year, the lord councillors and the entire Council of One Hundred went to make the vow in that church.[76] Three councillors walked in front and three behind in full regalia, and the council members walked in between them marching two by two, with the nobles and honored citizens[77] in one group and the merchants and artisans[78] in another. They marched in this order from the City Hall[79] to the church where mass was said amid much singing, and there they most solemnly swore the vow. There were so many people that neither the church nor the square outside was large enough to hold everyone and the Holy Sacrament was kept on display in the church.

Since the Saint's feast day was on Palm Sunday that year and the procession could not be held that day, [42r.] it was decided to postpone it until the Monday of the first week after Easter. This took place with much solemnity, leaving from the Cathedral and following the Street of Hell, then Saint John's Way in front of the Magdalenes convent[80] and straight to the church. There the procession was met by the brothers of the monastery to the great joy of all and with as large a crowd as was imaginable. The city also voted to make a silver statue of Saint Francis de Paul to carry in the procession. The mass was said there with great solemnity and fireworks were shot off when the Host was elevated. And all Barcelona was lit up during the three days of the Saint's feast day celebration, and there wasn't a house without candles in its windows, such was the eagerness and devotion with which everyone wished to serve the Saint. Everyone was glad that the vow had been made and that in the future the same would be done every year. And the city also ordered that a large painting be made with the figure of Saint Francis appearing in the heavens with a rod in his hand and the city underneath him as if he was defending it, along with the six councillors painted kneeling with their hands in prayer pleading for the city's health, and that this painting be carried to the church.[81] And if you wish to know who were the councillors that year you can look back to page 39 [of Parets's diary] and find out.[82]

❧ How They Carried Saint Madrona to the Cathedral for Rain and Against the Plague

Even though a few pages back you will find that on March 16 of this year they returned the body of glorious Saint Madrona because she had given us plenty of rain and because they had promised to do so, and even though at that time we had enough water, it pleased Our Lord to calm down the weather so much that it did not rain from early March to the end of April.[83] And since it was time for the wheat to sprout open, the drought frightened everyone. Seeing that things were already so bad, everyone feared that they would become even worse next year. And since the plague kept spreading, especially at the full moon of April, it caused great havoc and a great many people fell sick and many died since the plague was so rampant. If people had fled from Barcelona before, even more fled now, and with much greater difficulty than those who went earlier because of where they went and the quarantines they were obliged to undergo, as will be explained below. Since on the one hand there was a great drought in the countryside and on the other the plague kept spreading, it was decided to fetch the body of glorious Saint Madrona in order for her to intercede not only against the plague but also for the rain which we so badly needed. Wheat couldn't be found at any price, not even at 25 or 30 pounds a *quartera* from good friends. The city of Barcelona would have suffered greatly had it not been for two voyages made by a large ship built in the city two years ago. This ship made two trips to Livorno[84] and brought back nine or ten thousand *quarteras* of wheat each time, and were it not for this boat Barcelona would have suffered greatly. They brought down the body of glorious Saint Madrona on April 24, 1651, carrying her with great devotion, and a few days later Our Lord was pleased to give us some rain through the mediation of glorious Saint Madrona.

How the Cathedral Held a Procession with the Relic of Saint Severus for Relief from the Plague

The city of Barcelona, seeing that Our Lord was so angered with us and that the plague kept spreading, and that so many prayers and processions and devotions were not enough to placate the Lord's anger, decided to hold a procession and carry the relic of glorious Saint Severus[85] along the entire Corpus route.[86] This procession took place with great devotion on April 30, 1651, and was attended by the lord councillors and the governor, Don Joseph de Margarit i de Biure, Marquis of Aguilar,[87] and the wool weavers,[88] who marched with torches and dressed as pilgrims, as they always do whenever the relic of the glorious Saint is brought out. A great crowd came to this procession and since this disease does not favor large gatherings of people, it was decided to ban any other street processions.[89] Neither did anyone dare mix with anyone else in the churches. And the monks closed the chapels in all the monasteries, and when the parish clergy [42v.] came out of the sacristy to say mass they were not allowed to have any contact with the monks. Instead, the latter set up some gratings in front of the chapels. The priests stayed in the chapels and only the regulars could walk around in the church, and this was how it was done in the monasteries. One could walk from one chapel to another, and where they were not connected they set up a barrier in front of the main altar and all masses were said there. And this was done in the monasteries so that the monks wouldn't catch the disease. They also kept their doors shut and they wouldn't let any parish priest into the monasteries themselves.

❧ How the Sacraments Were Administered During the Plague

The plague kept spreading in late April and early May. The dead and the sick were now carried in heaps to the pesthouse of Jesus, and the vicars in charge of the parishes either fled or died. Since none of the priests wished to serve as vicars and administer the sacraments, monks were sent to all the parishes to administer them. The number of monks varied according to the size of the parish, and they ate and drank and lived in the vicarages and walked around with their robes cut short up to their knees.[90] If there were two in a parish one went in front confessing and the other carried the Holy Sacrament behind him. Each carried a torch, for when confessing the sick the torch was held between the priest and the sick person, and they kept their distance because it is said that the plague is carried by one's breath. Thus they stayed far apart and they did not spend a long time with the confessions. When giving Communion they extended the Holy Sacrament on the end of a silver rod in order not to touch the sick person, and they gave them Communion and the last rites at the same time in order not to have to return. After the Communion they gave the sacristan 8 shillings and sixpence in parish fees. This is how it was done in the parish of Saint Mary's, as I did not find out about any others.

When they went out to give the last rites they didn't do it just for one or two persons. Instead, people went to the vicarage and wrote down the name of the street and of the house, and since there were so many sick at that time the priests went around in turn to visit the sick and to give out food to everyone. In the parish of Saint Mary's,[91] many times when they went out at the height of the plague (the end of May, June, and early July) they gave Communion to seventy or eighty persons before returning home. Thus the poor monk was exhausted when he returned to the church from such a long walk and from having climbed so many stairs, for most of the sick were up in attics in order for them not to have contact with anyone save the person nursing them.[92] When giving out Communion, no priest came along, just the monk who carried the monstrance and the sacristan who carried the torch. They did not carry any canopy for the Host; instead the monk just carried the monstrance and its cloth, and it often happened that they went alone with no one behind them to accompany them. This was frightful and pitiful to see, and had it not been for the bell that the sacristan rang, no one would have known they were there. When they heard the bell, people had them come up, and the priest confessed them and afterwards gave them all food.

Thanks be to God, that in this sort of disease people do not die suddenly, and that one did not hear of anyone who, when feeling struck by it, did not quickly put his spiritual affairs in order. The sickness gave him time to do so and this is what all discreet people did. Just as quickly as they sent for doctors they also sent for confessors. Still, there were others who refused to take it seriously, and in order not to become frightened waited to see what the next day would bring. Thus by the time they wanted to confess they could not, because of the sudden turns and side effects of this sort of disease, and they died without the last rites. Yet one hardly heard of anyone who wanted the last rites who died without them, which was a great mercy of Our Lord. Many of the monks who administered the sacraments died because they ran great risks, and when one fell sick another came to take his place and just as cheerfully as if he was going to martyrdom.[93] God knows the good they did for so many souls who would have died without confessing had it not been for them, and they guided them to heaven and thus earned great merit while facing great dangers. Where there is risk, there lies the merit. When feeling sick each one went to his own monastery or to a building set aside as a pesthouse, and there they were given food. If they were cured they underwent quarantine and then went back to serve where they were assigned and where they were most needed, and as far as this was concerned Barcelona was well governed.

❧ The Flight of the *Diputats* and Royal Council and the Placing of the Stakes

Although the plague was raging in Barcelona, the city still hesitated officially to declare its presence out of fear of the great harm that would follow; it was very low in supplies of grain and other foodstuffs and people still hoped and waited for the sickness to subside and for Our Lord to have mercy on us. Yet as things continued to get worse and the plague kept spreading, there was no choice but to declare it such.[94] Once this was done, the *Diputats*[95] and Royal Council of Barcelona had to leave, because during times of plague they must stay outside in order better to oversee the care of the city. The decision was not made public until late April 1651, and before announcing it the *Diputats* and all the other officials went to the town of Terrassa.[96] And the judges of the civil chamber of the Royal Council dispersed towards different parts of Catalonia, while those of the criminal chamber stayed near the *torre pallaresa*,[97] which is near the monastery of Saint Jerome of the Myrtle not very far from Barcelona. Meanwhile, all the legal business of the Royal Council was postponed indefinitely.

As these lords left, Barcelona was emptied of a great many persons who, terrified by the plague, fled from the city. Previously they had had high hopes that all this would be patched up[98] and that it would be nothing and that everything would be taken care of before they would have to pack up their things and leave. This is what those who now fled Barcelona had thought, that is, they waited for matters somehow to straighten themselves out, but this should not be done during the plague; instead, one should always try to be among the first to flee. And those who left at this time suffered great travails and disasters and had to pay a high price, for all Catalonia had been alerted and so many roadblocks had been set up and so many guards were stationed in all the villages and towns and cities that not even a cat could get by. They refused to let anyone get by or even get near, nor did they allow anyone to enter a single house or anywhere. Instead, people had to sleep out in the open without shelter until they got to where they were heading, where they were made to stay in huts (although this was not true everywhere) and where they were closely watched by guards whom they themselves had to hire. They were made to undergo quarantine for forty days, and their clothes were burned and they were made to put on new clothes and to suffer even worse misfortunes, as will be explained at greater length below in the part about the sufferings of the plague.

Once they were out of the city of Barcelona, the lord *Diputats* and the Royal Council had stakes set up in order to encourage outsiders to bring the provisions the city needed. Thus towards the east near the New Gate they placed the stake at Saint Martin's Bridge.[99] Towards the west near Saint Anthony's Gate they put the stake near the slaughterhouse of Sants,[100] although they later moved it closer to the city because of the crowds of people who came from the surrounding area to the slaughterhouse. They put another stake on the sea side near the lagoon,[101] where boats came from hither and yon to bring supplies to the city. They unloaded there, and skiffs came from the Barcelona docks to carry the supplies to the city. There was no stake near Angel Gate, because no one was allowed to enter or leave through this gate except the town officials and grave diggers, who carried the sick and the dead to Jesus along with the provisions and rations for the sick, and it was forbidden to set up any other stakes under stiff penalties.

This was the way the stakes worked: they dug some wide and very long ditches in the middle of the highway. Those who came from outside the city stayed on the far side and brought their goods, while the city dwellers stayed on this [the city] side and brought their money. In the middle of the ditch three long planks were placed going from one side to the other, with a pole in the middle attached to a strong metal axle, which held up the middle plank and allowed it to turn around like a toy wheel. When the farmer brought his goods, chickens or eggs or fruit or anything else, he put them on the end of the plank and spun it around to the other side. If the buyer liked it they agreed on a price, and then the person from the city put the money on the plank and spun it again. And the farmer put the coins in a pot of vinegar he brought with him and afterwards he counted them, and in this way [43v.] supplies reached Barcelona. Retailers came out from the city to the stakes and brought vegetables and fruit to the

Born marketplace.[102] There they sold them to those who did not want to go to either the inland or seaside stakes, and thus almost nothing was lacking in the city.

The stakes were guarded in the following manner: the city placed two trustworthy men in charge of each stake, and gave them four or six soldiers with muskets from among those the city had paid to guard the walls, relieving them every twenty-four hours. And these soldiers were placed under the orders of the city's two trustworthy guards and took up position all around the ditches to make sure that no one from the city got near the people outside and that no one crossed from one side to the other, which was the sort of measure needed to be taken against the plague. Two huts were placed on the city side, one for the guards and the other for a victualler who sold refreshments to those who came to the stake. On the far side of the stake there was a large hut for the judge of the Royal Council and his personal guards, and each judge was to make sure that provisions were brought to the city. But these cared only about looking after themselves instead of the welfare of the city.[103]

The judges also made sure that there were guards on the far side of the stake so that there would not be any contact between either side, and to issue plague passes to those who came to the stake. And in this manner people could come from any part of Catalonia to the stakes and summon whomever they wanted from the city and speak with them and receive letters from wherever without running any risk. And once the stakes were set up no one could leave Barcelona and go beyond the stakes without permission and a pass from the councillors. And stiff penalties were decreed to ensure this. And those from outside who wanted to enter could not leave without permission, and wherever they went they were made to undergo rigorous quarantine. So their reasons for going into the city had to be very important. And those within the line of stakes could enter and leave the city without permission and could take to the Born whatever they wished, which some did, but the rest did not because they would be in great danger if found without a pass. Thus virtually no one cheated, for if they left the city they would not be allowed entry anywhere else if they did not carry a pass saying where they came from, unless they carried forged passes saying they were from elsewhere. If they were caught with these they were sure to get the death penalty, and thus everyone tried to behave themselves.

How Monsieur de Marsin Returned from France to Take Command of the Army

During this year 1651 France did not send a viceroy; instead, it once again sent Monsieur de Marsin[104] as military commander, and if you look back to page 31 at the beginning of 1650 you will see how this Marsin had been arrested in the city of Barcelona by order of the *parlement* of Paris; there you will find the reasons for the arrest of the Prince de Condé[105] (and as Marsin was of Condé's faction, he was arrested also). He was taken as a prisoner to the castle of Perpignan by Don Joseph de Margarit i de Biure, governor of Catalonia. After the Prince regained his earlier influence and position, M. de Marsin returned to office. Because the oath to the king of France was supposed to take place this year and the peers and ministers all wanted to be present [at the occasion], or for some other reason, they didn't send a viceroy to this city. Hence they sent Monsieur de Marsin as military commander. And thus he arrived at the end of May. And since the plague was raging so much in Barcelona he stayed in Granollers.[106] From there he came as close as the stake of Saint Martin's, where he was met by the lord governor and the councillors, and there they discussed important matters touching the war effort, for the enemy was assembling a large army near Lleida[107] [44r.] in order to place Barcelona under siege, which is indeed what happened. After discussing these matters he returned to Granollers and from there went with his staff to Cervera[108] to assemble his troops and arrange matters.

How All the Prisoners Escaped from the Royal Jail

The plague spread so much that it came to the prison, and since the prisoners are always looking for some way to escape, sometimes yelling the hue and cry hoping to incite a riot, now they started to shout, "Help! We are burning with plague, take us out of here!" And while things were not too bad there, they still exaggerated in order to frighten people from entering. And since a few really did die of plague and since they are so crowded in there and it was feared that it would get much worse as it became hotter, the guards did not want to be there nor did the charity officials at the Cathedral[109] want to take any food in there, because it was said that there was more plague in the prison patio than anywhere else in the city. They thus decided to run away, and it was said that the guards fled and left the keys with a housewife. I think that the law officers cared little about it. Seeing that people suffered so much from the plague and that all of them would die of this sickness if it worsened, they made little effort to guard them. Thus at the end of May, which was the week of Pentecost, the prisoners in the outer cells and the patio began to flee at midday. The lord governor was informed, and he immediately came with his people and locked the place up again, even though many had escaped, and even then they refused to add any new guards. And three days later the others began to escape; they got the keys one afternoon and those in the upper cells and the remaining prisoners in the outer cells and the patio all fled, everyone except the prisoners in the dungeon.[110] The governor gathered a large posse and sent it out to make it look like they were doing something about it, and they caught two of the escapees in front of Saint Catherine's and took them back to the prison. But since things had already begun in this way and were headed for a bad finish, and no one wanted to take food in there, the guards themselves opened up the dungeon and brought out the prisoners in chains because they were suffering so. Thus they all fled, leaving no one behind, not even those condemned to the galleys (of whom there were many). They all escaped, and thus the jail remained empty, without a single prisoner at this time, and they fumigated and cleaned it thoroughly before putting any new prisoners in there in order to remove the stench of plague that was still in there.

❧ How the Plague Continued in Barcelona in June and July

During the feast of Pentecost and the following week, which was toward the end of May and early June, the plague wreaked great slaughter in Barcelona, so much so that now the grave diggers who were divided among the city's six districts (one district per councillor) could not carry all the dead and sick to Jesus.[111] They had to use carts to carry the dead, and the grave diggers themselves carried the sick in cots. Each cart was accompanied by a deputy of the plague warden whose job was to keep people out of the streets when the carts passed by. It was terrifying to see the carts (there were one or two or however many were needed in each district). To see them move through the streets filled with the dead, some fully dressed and others naked, some wrapped in sheets and others with only their shifts on, was a terrifying sight, and there were so many of them that anyone going through the streets constantly ran into them. Other carts went about carrying the mattresses and blankets and sheets from the beds of the dead. This was done by the grave diggers, who carried the sick and removed the clothing from the beds where the sick had lain and took it to the pesthouse, which had gotten so crowded that a sick person who did not bring along a mattress and bed clothing had to lie on the ground. So many sick people went there that not only were there not enough beds, but there was not even enough space in the entire monastery of Jesus for them. They had to make some shelters out of wood in the garden in order to find a place for the sick, as there was not enough room for them all upstairs or downstairs, and at one point the number of sick reached three or four thousand. Even though a great many died, there were so many who suffered greatly from lack of food because there wasn't enough for so many people.

And seeing that things were going so badly, many people got thoroughly frightened and fled the city to live in huts, some in Valldonzella[112] and some in Montjuich and some [44v.] around Sarrià and others even further away, resigning themselves to having to undergo quarantine wherever they went. These people suffered a great deal, as will be explained below. Thus the city was left with very few people, and it was terrible and piteous to walk through the streets, as one hardly ran into anyone else except someone looking for food for the sick.[113] The provision of food during the plague was badly organized, both in the pesthouse and in the city.[114] Many died in the pesthouse for lack of food who would not have died had things been better organized, and there were many who after arriving in the pesthouse died without having been seen by a doctor or surgeon. And all the doctors and surgeons in the pesthouse were very young fellows without much experience, because it was so hard to find doctors willing to go there, especially after some of the first ones died. In the city there were a few good doctors and some came from Girona and Olot who had some experience with the disease, and the city government gave 10 pounds to each of them, surgeons as well as physicians. And they earned even more from the payments they received from private citizens.

When the city saw that the plague was getting worse and that it kept spreading from one person to another, it ordered under penalty of death that no man or woman who had nursed a sick person could move about within the city nor could anyone have any contact with those nursing the sick. The city then hired people in each quarter to supply them with meat and soup and whatever they needed, and prohibited anyone from entering any house where there were sick persons. It also ordered the doors of the houses of the sick to be marked with the white cross of Saint Eulalia, so that when people saw the cross they wouldn't enter. They also ordered the houses whose occupants had died or had been taken to the pesthouse to be nailed shut, and that no one dare enter without permission from the councillor of that district under penalty of death.[115] In this fashion things got better.

❧ Of the Punishment of a Thief in Front of the Marcús Chapel[116]

During the plague in Barcelona, those left in their houses after the other inhabitants died or went to the pesthouse who could not take care of themselves or were orphans without anyone to look after them were taken to some houses on Jesus Street called the "purge." There the city fed them or had them taken to the university building or the seminary near the Nazareth Church.[117] There they were made to undergo quarantine and a purge[118] and, if necessary, they were made to undergo yet another purge to make sure they were clean. The plague officials had orders to close all the houses thus emptied in their district and to nail them tightly shut with a piece of tin on the outside, and the death penalty was decreed for anyone entering these houses to rob or even to touch anything in them before they were well cleaned and fumigated. There were many scoundrels (as there always are on these occasions) who broke into these houses at night after making sure that no one was in them and stole the best stuff within. To combat this the city government named a marshall to go after such persons because of the great harm they were causing. At first the city did not wish to be too harsh, so they only whipped the ones they caught and sent them to undergo quarantine or to work in the pest-house of Jesus. In early June a youth who was a servant of the royal courier,[119] Don Francisco de Aiguaviva, broke into the house of a druggist named Joseph Vinyes, whose shop on the corner in front of the Marcús Chapel had been closed. This was because of the death of a brother of Vinyes, who had been left in charge of the shop while the owner and his family were outside Barcelona. The culprit was captured an hour after robbing the house, and the lord councillors, who have absolute legal authority in time of plague,[120] condemned him to a hundred lashes and spared his life on the condition that he serve in the pesthouse. He was taken there, and since he obviously was a man fated for the gallows, he fled from the pesthouse as soon as he could and returned to Barcelona and went back to rob the same house. They immediately caught him and condemned him to death on two counts, first for having robbed a plague-stricken house, and second for having left [45r.] the pesthouse and entered Barcelona, which was punishable by death. Thus he was condemned to hang, and to set an example to other evildoers, they set up a gallows in front of the house and hanged him there. This sentence was just what was needed to solve the problem of houses being robbed.

❧ Another Sentence Carried Out in the Pesthouse of Jesus

Three men were also ordered garrotted in the pesthouse of Jesus for having broken the orders prohibiting persons from going from one place to another, that is, no one could go from Jesus to the purge houses on the same street, neither the sick nor the persons looking after them. Neither could they enter Barcelona without permission, and those who entered with permission, whether surgeons or the others serving in the pesthouse, had to go with a guard to make sure that they had no contact with anyone and to warn people in the streets not to get near them, even though the doctors and surgeons wore a white cloth sash and were not allowed in without it. And since there were some who still made it their business to go from one place to another, three of them were ordered garrotted. This was also because of the great crime they committed in the pesthouse of having contact with the women there. It was truly said that there was great lewdness in the pesthouse regarding women, that it was like a little brothel,[121] and that when a good-looking woman who was brought there caught the eye of one of the orderlies she was well treated and given special care, and when she was well they took her aside and kept her for their own use while she was undergoing quarantine and they kept her there for as long as they wanted. And this went on so long that it caused great fear and terror in whomever heard it, that in a place where one had death constantly before one's eyes and seeing the anger of God so mightily upon us, and that there, where one should be most pure and most fervently committed to God because one ran the greatest risk of death, there Our Lord was most offended. When the councillors heard about this they ordered such great knavery and abuses to be punished. This was done, and copious lashes were given to the women caught breaking the law in the pesthouse. As for the men, if they were orderlies they were removed from Jesus and their jobs taken away and they were forced to undergo quarantine. And it was said that in the pesthouse only bad women had good food, and even if they came there virtuous and with a good reputation the very orderlies and persons looking after them made them bad, but thanks to this punishment things got better.

Of the Great Sufferings the Plague Caused

The physician Dr. Rossell[122] says towards the end of the book he wrote during the last time plague struck Barcelona that there is no better remedy for the plague, no matter where one is, than to be among the first and farthest to flee and to be among the last to return when the plague has been long forgotten,[123] and in this way you will escape this sickness. I say that it is quite right to flee in order not to suffer from this disease, for it is most cruel, but it is just as right to flee in order not to witness the travails and misfortunes and privations that are suffered wherever the plague is found, which are more than any person can stand. One saw that whenever anyone fell sick he lost all touch with friends and relatives, as there was no one who would risk contact with him, just the person nursing him.[124] And it would have to be someone very close or related to him who would dare to take care of him, like a wife to a husband or a mother to a son or a sister to a brother, and even among these many fled or did not want to stay, for the plague was so evil and of such a bad sort that everyone fled.

I can give good account of all this, because my wife fell ill with a plague boil on her leg and another on her thigh, from which she died. And although she had two sisters in Barcelona, neither was willing to come to nurse her, I mean not to nurse her but even to see her, for they could have seen her without having to come into my house, as they could do so from the house in front where everyone was quite well and healthy and from so close they could have seen and talked with her. Although the sick woman sent for them because she wanted to talk with them and see them before she died, there was no way to convince them to come, as everyone fled from the plague. And there were many cases of this sort.

The sick had to find someone, man or woman, who would nurse them for pay.[125] These persons received 12 or 14 *reals*, along with expenses for each day they took care of the sick, and they also arranged for the payment of their quarantine after the sick person was either cured [45v.] or dead, for which they charged 18 or 20 pounds and some even more. But they did not undergo quarantine at all, as the demand for nurses was so great that when they left one house they went to another, and even then they were hard to find. Those who did not find anyone this way had to hire women or men who were undergoing quarantine at Jesus who had already had the plague and thus were not that afraid. This had to be done with the written permission of a councillor allowing these persons to leave the quarantine, and all this could be accomplished only by paying money.

Just think about a person who, during other sicknesses, was taken care of by his or her spouse, and was able to see his relatives and friends, and was given everything he needed. And then you see the same person during the plague being nursed by a stranger with no love for him, or perhaps never seen or known by him before, and he had to receive everything from this person without being consoled by any other. And many times all this nurse did was to make the patient die more quickly, because the sooner he died the sooner the nurse got the 18 or 20 pounds or however much they had agreed on for the quarantine, and then the nurse would be free to go elsewhere. Many times those taking care of the sick did not give them the medicine that had been prescribed, nor the food and soups they were supposed to feed them, and even if the sick were supposed to eat the meaty part of the soup they were given only the watery part. Since there was no love or acquaintance between them the nurses did not bother to take good care of the sick; instead they looked after themselves. This was seen in many cases and many of the sick died from vexation and despair over these very things.

Now let us consider how the poor babies who were still nursing suffered,[126] for when their mothers fell sick they no longer had any milk or were otherwise prevented from nursing with their mothers. But for those babies whose mothers died and for whom wet nurses had to be found, they could not find anyone willing to nurse them. May good Christians never feel the pain of the fathers who went from door to door looking for wet nurses or milk for their babies.[127] No one wanted to help them after the child had nursed with a mother or another wet nurse who had fallen sick, because everyone was afraid of catching the plague, and even if one somehow found a wet nurse, it was thanks to a large fee, 100 or 150 pounds, depending on the ability of the woman to feed the baby. But before the child was allowed to enter the wet nurse's house he was stripped naked and washed in vinegar and rubbed with lavender and other soothing herbs and passed over the flames in the fireplace, which was a pitiful sight to see. Afterwards they had to find fresh clothes for him, as he could not wear anything he had on before. Worst of all was when, after taking all these precautions and the child had settled in, within a few days the wet nurse herself would fall ill. Then they had to do it all over again and find another wet nurse, which was a great vexation for his poor parents. And if the baby fell sick while with the wet nurse, the nurse returned him to his parents, who would not be able to find another wet nurse for him now at any price. When this happened they had no choice but to send him to one of the houses on Jesus Street, to the wet nurses the city had hired for this purpose by giving them huge salaries and providing them with everything. The babies who were sent there were labelled with a ribbon or tag around their arms or leg or neck with the names of their parents written on it, so that the parents would know them if they ever got out, but hardly any of the babies sent there to nurse managed to survive because so many were sent there. They found few wet nurses willing to go there, as every day they had five or six babies to take care of, which were too many to clean and feed. It was said that it was the greatest pity to hear the crying from the wet nurses' house, that it sounded like a goat pen. And since the wet nurses at the hospitals[128] are normally like cows, and are evil and uncaring women who would rather the children die than to live, they let them cry and did not bother to change or to clean them, and the poor babies were peeling all over.[129] Thus many of them died, either from the plague or from the wet nurses' bad care.

There were many mothers who, when they fell sick and couldn't take care of themselves at home, were taken with their babies to the hospital, with the mother going to Jesus pesthouse and the baby to the wet nurses. And if there were a great number of children without nurses, they were taken to quarantine [46r.] if they were well and if sick they stayed with the sick. Some of those in quarantine managed to survive and returned to Barcelona clean and purged and having undergone the entire quarantine. There were many who, when taken to Jesus had fathers and mothers, and when they returned did not have anyone, and almost no one wanted to take them in out of fear of something bad happening. For this reason many children wandered about lost, which was the most terrible thing in the world.

We should also think about the sufferings of the mothers or wet nurses whose baby died of plague and afterwards their milk greatly bothered them. And when it was known that their child had died of plague they could not find anyone willing to give them a baby to nurse and their milk hurt them a great deal. The only babies they could find were those whose mothers had died of plague, and this way well persons came into contact with the plague and risked catching it from others. Sometimes they lived and sometimes they died, the one or the other. It was God's will, but the poor little ones suffered greatly and a great many died.

We should also think of the poor women who were pregnant at this time, among whom perhaps two among every hundred survived.[130] If they were in the last days of their term all they could do was to commend themselves to God, as most of them simply gave birth and died and many of their babies died with them. The midwives refused to work out of fear of catching the plague, so other women had to serve as midwives, as did many husbands. And if the baby lived, his father could not find anyone to cradle or to nurse him, and he walked about with the baby in his arms looking for a wet nurse or someone who would help them out when the child needed some milk. Often they found a wet nurse in the way noted above, one whose baby had died of plague or who herself had been sick, and in order to have someone to nurse the child they had to take their chances with whomever they found.

Among the greatest pities and heartbreaks that I saw and suffered myself was to see someone die without being attended by a priest to help him or her remember the Passion. The only one who could help was the person taking care of the patient, regardless whether it was a man or woman. Often it was a Frenchman, God knows whether a true Christian, because there were many Frenchmen taking care of the sick who perhaps instead of helping them die well were in a hurry for them to die.[131] If the sick person was a woman and wearing rings on her fingers they took them, sometimes even pulling them off without waiting for her to die. Other times the nurses wrapped the sick up in a sheet and they died with their head or feet tied down, and until the grave diggers came to take the corpse to Jesus the nurse was alone in the house, and God knows the pillaging he did since no one dared enter and he could do whatever he wanted. And when the nurse finally left the house because there was no one else there who had to undergo quarantine, they nailed the door shut and would not open it until after the plague was over, when they fumigated the houses in the way noted at greater length below.

Now let us consider the sufferings of the people who, while fleeing the plague, left Barcelona and gathered in huts on the slope of Montjuich mountain and in the plain of Valldonzella and other places not far from Barcelona.[132] Since the plague was raging so much, they could not go very far away because, regardless of quarantine or guards, when it was known that they came from Barcelona no one would give them any shelter. Thus the last persons to flee had to stay within the area of a league around Barcelona, and even then only close relations or friends or persons greatly obliged to them would take them in, for few were willing to run the great risk of exposing their entire household to plague. Even then there was great reluctance to let them stay in huts, and many who had no acquaintances or friends where they could undergo quarantine outside Barcelona simply had to be patient and wait inside Barcelona in their houses, making sure to do whatever they could to not go about the city by having someone bring whatever food they needed to their houses. They would not let these persons enter nor have contact with anyone inside, and if what they brought was meat, they boiled it first, and if it was something else they fumigated it or washed it in vinegar. There were many who protected themselves in this way. But many of these, without thinking and not knowing what to do or perhaps out of their great fear of the plague, for they were always full of fear, when someone fell sick they completely abandoned their house [46v.], and those who were well had to find someplace else to live and they could not find anyone who would shelter them, for everyone feared the same and shunned everyone else in a terrible fashion. And they had to find someone to take care of the sick who stayed behind in the house, or they had to have them taken to Jesus if they could not nurse them at home.

There were also others who did not have anyone outside to shelter them or let them stay in a hut. And they were so afraid and so tired of seeing and hearing so many misfortunes that in order not to be within the city nor to see so many travails, they left to live in huts either on Montjuich mountain or the plain of Valldonzella, taking their families with them and making huts out of dirt and sticks or of timber and branches. But this way of fleeing did them little good, for they ran the same risk, except that the air outside was better and more healthy. Neither did they continuously witness as many deaths, as many travails, as many privations as they had in Barcelona, which was enough to break the hearts of stones, much less of people. Thus those who were outside Barcelona in the area around its walls were only far enough away not to see and hear all these things, but for the rest they were in as much danger of the plague as those inside the city, because they had to have a person enter into the city one or two times a week to bring them the food they needed, for even though at the beginning they brought many provisions, these were soon used up. Thus it was necessary for someone to go inside to find supplies, and either this person had to have contact with the people in the hut or the provisions from the city could carry the plague to them. And it goes without saying that they could hardly go to look for supplies in Sarrià or Hospitalet or Sant Andreu,[133] because even though these places had their own problems with the plague they were very suspicious of anyone from Barcelona. Thus there was no choice but to go to Barcelona to look for provisions. People took care to not have any contact with those living in nearby huts, as many within them fell sick and caught the plague and then they had to be taken to the city to be nursed or to Jesus. When this happened the whole family would have to move, as they had to burn the hut and return to the city to face the same danger as before. And a great many among them wanted to stay to take care of the sick in the huts in order not to have to return to the city, and many died in them and many without having confessed as Christians and they were buried in the fields.

There were others who had relatives or good friends in villages around Barcelona like Sarrià, Sant Andreu, and Horta[134] and Sants (etc.). There they built one or two good shelters, depending on the size of their family, and were able to keep clean and washed, and change or burn their clothes if need be. And they built their huts near the houses of their friends and relatives, and those within the houses sent them food, while making sure that they did not enter into contact with them by leaving the food outside the huts, and then those in the huts would go fetch it. And the other villagers made those who lived in the huts keep one or two men as guards at a line marking a boundary so that they would not have any contact with anyone in the house nor would they wander from the huts. The people in the huts paid these guards, giving them 10 or 12 *reals* every day. These people were well off because they didn't have to have contact with anyone from Barcelona, nor did they have to get their supplies from there. And after thirty or forty days of keeping quarantine they changed clothes and fumigated and washed themselves with vinegar and went to stay in the house of the friend or relative who had brought them food.

Many of these persons managed to survive. But some of them also died, and when they fell ill they suffered twice as much as the sick in Jesus or in Barcelona, for when they became ill either a hut was built for them far away from the others and a person was found to nurse them, or someone from their company was obliged to nurse them. A doctor had to come from Jesus or from Barcelona and a surgeon as well, and all this cost a great deal of money, and they were very slow in getting there. They had to hire someone to look for medicine in Barcelona, which cost a lot of money. They could not keep them comfortable nor could they look after them the way they should have, nor with the diligence and care that they needed. They had to take their relief as they found it, and many times after the plague had already done its worst it kept on killing because the sick lacked food. In this way the plague [47r.] kept raging, and many died without confessing as Christians and afterwards were buried in a ditch. And others who had nothing left to spend died for lack of food, for they found no one who would give them anything. Many poor people fell sick while walking the roads and kept walking as well as they could, and when they could not walk any more they lay down in a ditch gasping for breath until they died. Even when people passed by they all fled, for no one would dare get near enough to say anything to them or to give them anything. It was such an evil sickness that it allowed little charity, so little that people fled and shunned not only the sick but also the well, as no one dared to get near anyone else for fear that they might carry the plague stuck in their clothes, which was so often the case.

And those who were among the last to get to the huts on the outskirts of Barcelona suffered a great deal. Well or sick, people were better off in the city; the sick there were better off than those who fell sick outside, and those who were healthy inside also did better than those who were healthy outside. I say this because of the supplies that could be bought with money in the city for both the well and the sick. Of all this I can give good account, after having gone through so much suffering from the plague within Barcelona, with my wife dying and a son almost thirteen years old and another who was eleven and a little girl one year old whom my wife was nursing, four persons in all, all dying in less than a month. My mother[135] and a son four years old also caught the plague and recovered, and all these dead and sick persons were cared for in my house and I can give good account of what these sicknesses cost. And when I saw so much misfortune I wanted to leave the city to see if I could escape with my life. And thus a brother-in-law, my wife's brother named Benet Mans, a peasant from Sarrià,[136] sent for me and my four-year-old son who was cured of the plague, and built us a hut next to his house and brought us food. And we left Barcelona on June 9, 1651, which was the day after Corpus that year. And we stayed in the hut twenty-seven days and afterwards we were allowed into the house, where we stayed until August 4 of that year when the Castilians came to besiege Barcelona.[137] During the time that I was outside the city I witnessed the many travails of many persons who had left the city, so many that it would take a long time to tell it, but you should be able to imagine it thanks to what little I have already written.

What of the misfortunes of the sick who went insane?[138] From time to time one saw some of the plague-stricken who had gone mad run through the city, for there were many who went crazy from this disease. If they could escape from bed or from those nursing them, they went out into the streets completely naked or however they were dressed at the time and no one dared stop them. Instead, everyone fled them unless they could find some grave diggers or the nurse in charge of them, who made them return home. But often they were so strong that it was not easy to hold them down, and no one could help them, instead everyone ran away. It was a pity to see how some who, if they could get to the window, suddenly threw themselves to the street and died. Unlike the other sick, who could be watched by a single person, man or woman, the frantic were so strong that they could not be kept down, and in this way many of them died. Few survived among those who went mad. And some who went insane and were taken to Jesus had to be tied to the stretchers for the dead or else they would escape. And they screamed all the way there. And it was a pity to see that once in Jesus they ran about the wards and went to drink from the well there and after doing so nearly all of them died.

There were so many sick in the pesthouse that at one point they numbered more than four thousand. Those in charge were not able to keep count of them; meanwhile, the food supplies were so badly organized, as noted above, that at one point the sick would be there ten or twelve days without being attended, either for lack of food and medicines or for other reasons. During this time a great number of persons died, so many that it is known for certain that more people died in Jesus of bad organization and poor food than of the plague itself. Those in charge did things so badly that of all the sick who came there, [47v.] only bad women and the worst sort of people were treated well. If a good-looking woman who came there wished to protect her virtue, she was given very bad food and the soup they gave her tasted like waste water, while bad women were treated well. In this way Our Lord punished us, although not as much as we deserved, for truly it was a pity to see the lewdness of these women, especially the bad ones, who instead of correcting their ways were even worse when the plague was over. I think that the plague brought this, or the misery of our times, that not only the sinful women did not get better but also that many who were good turned bad.

When God wished the plague to get better in Barcelona, the few persons left alive had all been impoverished, because during these times hardly anyone thought of working, just of spending and of feeding themselves with the little food that God had left them. One cannot help notice that those who had more were able to make it last longer, while those who had little soon spent all they had and were forced to go about begging. Many people who had been able to get by spent whatever they had on illnesses, which cost God knows how much, as things became more expensive than one can say. People resigned themselves to spending whatever they had if only they could save their lives, and if they had gold or silver or copper they sold it all in order to be nursed and fed. Thus their savings were used up, and many people became impoverished. And what was even more pitiable was to see so many children left without fathers or mothers or any relatives when just a short while before they were so well off in their homes with their parents. And afterwards to see them lost and wandering about the city begging was a pity for those who knew them, and in this fashion many children fell sick and never grew up. Alas, there are so many pitiful things to tell of the sufferings caused by this plague that it would be a tale without end. Let every person consider what has already been said. And may Our Lord not allow Christians to witness such misfortunes, for the best thing is to go away and far away in order not to see so many travails and so many misfortunes.

That many of Parets's comments on the sufferings caused by the plague were autobiographical in character is made clear by the following passage, the anguished climax of his chronicle, which recounts the death of his wife and three of their four children.

[I, 141r.] On May 15, 1651, it was the will of Our Lord that my wife Elisabet Parets y Mans[139] die of the plague. She was buried the same day in Saint Mary's by the Sea. And the most that could be done for her was to have one mourner and four candles, for it was impossible to find wax at that time and burials cost a great deal. And she told me herself before she died that after her death she wanted the masses of Saint Vincent Ferrer and a service of Our Lady sung for her in Saint Catherine's Monastery,[140] which I had said for her right after her death. I paid 9 pounds and 12 shillings for the forty-eight masses, and for the service of Our Lady I paid 2 pounds. She did not leave a will, because at that time one could not find any notary willing to take down wills, thanks to the plague.[141] And thus she told me orally that of the 400 pounds that she had brought me as her dowry, 200 pounds should go for the marriage of Anna Maria (mentioned below). She also left 50 pounds to Miquelo and another 50 to Gabrielo, who were still living then. The remaining 100 were for her soul and mine and all others for which she was responsible.

It was the will of Our Lord that on the day after the death of my wife my one-year-old daughter Anna Maria died.[142] It was not certain that she died of the plague; she had been given to a young woman to nurse while my wife took care of Gabrielo,[143] who had a plague sore under his left underarm and later got well. And the day on which the boy began quarantine, my wife fell ill with a plague boil on her thigh and another under her loins, from which she died seven days later. And since the little girl either got tired of milk or could not nurse any more, the day before her mother died she fell ill with an intestinal sickness which lasted until she died. No boil or any other plague symptom was found on her. She was buried in the pesthouse of Jesus, since at that time virtually no one was buried anywhere else.

The second day of Pentecost, which was May 29, 1651, it was the will of Our Lord that the above-mentioned Miquelo,[144] my son twelve years and two months old, die of plague after getting a swelling in his throat.[145] There was no cure for it, and as soon as the doctor saw it he immediately gave him up for dead. His sickness only lasted a day and a half, and when he died it was as if he suddenly choked. He was the oldest son. He was buried in the pesthouse of Jesus.

And on May 1, 1651, [our son] Josep died.[146] He was ten years and eight and a half months old, and died of a disease of a bone in his arm following five years of illness which cost me a great many ducats, and during the last year and a half surgeons came daily to look after him. And he was so dried up when he died that all that was left were bones and nerves. One would never see a human body more dried up. He showed such patience when dying that he seemed to be a martyr. Three days before he died he called me and his mother to the room where he slept to show us Our Lady accompanied by many other virgins and Saint Joseph his patron saint and the Archangel Saint Michael and many other angels. He said that he noted a great smell of roses and other pleasing odors, even though there was nothing in our room. He also said that he heard wonderful music, and asked to be allowed to listen to the wonderful music, which made his mother leave the room. He was fortunate to have seen all this for we [141v.]

didn't see anything . . . [illegible] . . . he often said that he saw these visions . . . [illegible] . . . buried on the second of the month in Saint Mary's by the Sea here in Barcelona. May he intercede with God for us. He was a very lovely boy and very comely and the best-tempered of all our children, very well-behaved and quiet and pacific. Despite all his suffering he always talked and liked to read, and one could tell from looking at him that God was calling him and that he was not long for this earth.[147] God took him at an age when he was very clean and very pure. And despite all that he suffered, when he saw his mother so afflicted and troubled by seeing him suffer so, he himself consoled her, saying that she should not be troubled, and that it was the will of Our Lord that he suffer such pain out of love for Him, and that when he got to heaven he would remember her.

This he surely did, for fifteen days after he died he called her to glory, after preparing for her the place and seat in which she would enjoy the presence of God. Of this one can be quite certain, thanks to her exemplary goodness and virtue, which will be vouched for by anyone who knew her well and spoke with her. In particular Brother Thomas Ros of the Dominican order, who was her confessor for many years, told me the day she died that all of us could envy her her death, because he knew her goodness and virtue and that she was surely in heaven. One could also be sure that she was in glory thanks to this sign on the day of her death: after breakfast, at ten o'clock in the morning she had her clothes changed and put on one of the best shifts she had, and after the bedding was changed she had the nurse taking care of her call me (because during the plague sick persons didn't have any contact with anyone except the persons nursing them). She had me climb up to the attic of the house next door, as she too was in a bed in a room in our attic, and from there I could see and hear her. She told me to give the woman taking care of her a piece of the candle of Our Lady from the Easter Matins, which I would find in the dresser, and the blessed candle of the Rosary, which I would find near the cot, and the painted wooden crucifix which we had over the bed in our room below, and she spoke so well and so clearly that it seemed that she wasn't at all sick. She then commended the children to me, saying that I should bring them up to be good and virtuous, and then specially commended to me our little girl, although she said that we would not have to look after her long. She then said good-bye to me with many tears. I never thought that she would die, and I went down and gave everything she asked for to the woman looking after her, and when I went back up there she took the crucifix in her hands and she prepared herself for death, for during the plague priests could not be found to help the sick die properly. Instead, that was left to those who nursed them or the patients themselves. And thus as I say she prepared herself for the end, which greatly surprised the woman taking care of her, and when she knew that it was time for her to die she had the candles lit, and with the candles and crucifix in her hands she died between twelve and one in the afternoon, Saint Isidore's Day, 1651. May God look after her in heaven.

And as I have written above, God took our little girl the day after her mother's death. She was like an angel, with a doll's face, comely, cheerful, pacific, and quiet, who made everyone who knew her fall in love with her. And afterwards, within fifteen days, God took our older boy, who already worked and was a good sailor and who was to be my support when I grew older, but this was not up to me but to God who chose to take them all. God knows why He does what He does, He knows what is best for us. His will be done. Thus in less than a month there died my wife, our two older sons, and our little daughter. And I remained with four-year-old Gabrielo,[148] who of them all had the most difficult character. And after all this was over I went with the boy in the midst of the great flight from the plague to Sarrià to the house of my mother-in-law. I kept quarantine there for almost two months, first in a hut and then in the house, and would not have returned so soon had it not been for the siege of Barcelona by the Castilian soldiers, which began in early August 1651, as will be mentioned in the other volume.

As the plague decreased in intensity, Barcelona faced a new threat—war.

❧ [II, 48r.] How the Castilians Came to Besiege Barcelona and How the Plague Ended

What Our Lord now worked in the city of Barcelona was a most miraculous thing and a great prodigy, that with the plague raging so furiously as has already been explained, and during the month of July which is in the middle of the great heat, it was the will of Our Lord to relieve the plague in Barcelona. Stopping the plague at this time and during the heat was not so great a miracle because there was hardly anyone left in Barcelona. One rarely came across anyone while walking through the most frequently travelled streets in Barcelona, for more than thirty thousand persons had died, and some thought that it was as many as thirty-six thousand, which was a most frightful thing.[149] In fact, the prodigy was not so much that the plague stopped at this time but rather that since the enemy army was coming to besiege Barcelona, everyone in the nearby towns and villages suddenly had to come into the city despite the plague in order to protect their grain and goods and whatever they could.[150] Everyone in the outskirts of the city came back inside, and it looked like the plague would spread even more as it usually did with the entry of new people, just like when one puts fresh wood on a fire. At this point Our Lord willed that the more people came in, the more quickly the plague subsided and came to an end, something which encouraged those who had fled far away to return to defend their homeland. Thus when the people who had fled saw that the plague was getting better everyone returned to Barcelona, that is those who were good Catalans[151] and who wished to defend their homeland,[152] those not thinking about what the plague might do ignored all possible obstacles in order to defend their homeland.

There were some perverse persons of wicked intentions, not worthy of calling themselves Catalans, who not only refused to come back to Barcelona, but instead if they were close by fled even farther. Seeing this, the government of Barcelona decided, in order to make its citizens return and defend it, that all those not returning to the city within fifteen days could not partake in the offices and emoluments of the city. If they were already eligible for the lottery for public offices,[153] then they would be declared ineligible, and if they were presently holding office then others would be chosen to replace them, while those who were not eligible for the lottery would not be allowed to be considered for it during the next ten years. During this space of time few came into the city, so an extension of eight days was granted, and when this expired they were given another eight days, so that three times the government ordered its citizens to come to its defense. Like a pious mother [48v.] she did not wish to punish her children, but as some of these were perverse and disobedient, the city had to execute its orders. These entered into effect on August 26, 1651, and in the following manner.

The government ordered some people to go throughout the city to write down the names of those who were absent and had not returned, putting them all into a single list which was read aloud before the entire council on the same day of the 26th.[154] After doing this, and since there were many who sent in petitions and depositions with justifications like sickness or other good reasons, these petitions and pleas were referred to a committee of twelve persons, who were charged along with the city's lawyers[155] with examining them under oath to see if they were justified. Afterwards, before the council adjourned, they read out the names of those whose petitions had been accepted and those which had been rejected, and then on the spot they removed from office some of those who had not returned. Since it was so late, with the council still in session at two o'clock in the morning, it was decided to charge the committee with continuing the next day to declare ineligible all those who had not returned. This was so done, although only after several meetings, including hearing the appeals of some who had returned late and wanted to explain why they had not come back earlier. Some of these were admitted and others not. It was also the will of Our Lord to punish some of these perverse persons who did not wish to come to the defense of their city by having them die outside of it. Many citizens died in this fashion in Girona and other places. The city did not declare ineligible for office anyone who did not deserve it, and in this it acted quite properly, for on the day in which it was done the enemy army had already been within sight of Barcelona for twenty-four days, and considering such great disobedience the city acted quite rightly.[156]

The First Thanks Given in Barcelona to Our Lord for Ending the Plague

The city of Barcelona, seeing the great prodigy or miracle of the end of the plague as has been noted above, decided in a meeting of the Council of One Hundred, held on August 6, 1651, that on the following day, that is the 7th of the same month, the lord councillors should go to the Cathedral accompanied by a host of citizens. There they should have sung a *Te Deum Laudamus*, and after the mass they should hold a procession within the church to give thanks to Our Lord for the favor of having appeased the anger of His divine punishment of the city. While it is true that the plague was still not completely over with, hardly anyone was dying or falling sick and only a few were brought to the pesthouse. And while a few people still fell sick, the plague was not as powerful as it had been and many now recovered, so many that in September the pesthouse of Jesus was closed. Those falling sick afterwards who had to be removed were taken to the quarantine houses on Jesus Street, where they were given care.

The city also undertook to have its streets and squares cleaned, removing the great filth that was there, for at nighttime people threw into the streets mattresses, blankets, sheets, and pallets and many other goods belonging to people who had died of the plague. The city had carts in all the districts and they went along collecting these goods, and they took them out past the walls and burned them all, and it was a pity to see all these good belongings burn. The city also had some huge ovens set up in the *fusina* near the Street of the Holy Spirit,[157] and they baked in these ovens clothes from houses where people had died. They hired for this job some persons to work there full-time, and after putting the clothes in the ovens they washed them and thus removed all the plague from them. Each councillor also had in his district some men called "fumigators"[158] who cleaned the houses, and those which had been shut and nailed up as described above could not be opened by their owners or anyone else without permission from the councillor of that district. Before anyone could enter, the fumigators had to clean them from top to bottom, and they had to wash the goods and beds in the rooms where anyone had died. And their owners could not come in until everything was clean, and this cleaning cost them a good many ducats. Thanks to these and other measures the plague subsided a great deal, but even so this was not thanks to these measures but rather to the mercy of God and to the pleas and prayers of some holy and good persons to whom Our Lord was pleased to listen. Truly in many things the right policies were not carried out, particularly in regard to the bread supply, because when we were under siege by the enemy people couldn't find grain to grind. Instead, nearly everyone had to go to the Customs House to get bread, and so many went there that it looked like it would burst, and what with the city meat shop right there and with the great crowds of people in the midst of the August heat, [49r.] the fact that the plague did not start up again was a sign that God wanted it so, and it is to Him that proper credit and thanks should be given. May He grant us the favor of the plague continuing to improve and that Christians never see it again.

Of the Vow the City Made to Present Silver Keys to Our Lady of the Immaculate Conception[159]

It was in late July 1651 when the Castilians marched toward Barcelona, having already passed Vilafranca[160] and coming straight for Barcelona in a great fury to place the city under siege. And when enemies come to besiege a city the first thing they usually do is to send a herald to call upon it to surrender and to hand over its keys. The arrogance of the Castilians is so great that they came with very few troops to besiege Barcelona, and it was said that they had only 11,000 infantrymen and 2,500 cavalry, which was very little for Barcelona. Since the city thought that they would send a herald, as noted above, it prepared its reply in advance. This was that the most wise Council of One Hundred would present some keys to Our Lady of the Immaculate Conception in the Cathedral, and if the enemy came to ask for them then they would receive the answer that Our Lady held the keys of the city and if she wanted to give them to them she would do so, and if not then they should be patient. She was thus very solemnly presented some keys made of iron on July 19 and afterwards four silver keys were made and given to her on August 25, Saint Louis' Day, when the lord councillors attended mass in the Cathedral. And when the service was over the lord councillors brought them to her with great fanfare, and after a *Salve Regina* was sung they placed the four silver keys in her hands. And the enemy must have found out about this as they did not send anyone to ask for the keys, and even though they began the siege a long time ago they didn't ever send for them.

Parets obviously relished this story, for he went on to repeat the tale of Barcelona's defiant presentation of its keys to its patroness, Our Lady of the Immaculate Conception. The second time, however, he added some new details, including the commissioning of a votive painting representing the councillors supplicating the Virgin's help against the city's enemies.

❧ [51r.] Of Some Silver Keys to the Gates of the City Presented to Our Lady of the Immaculate Conception

Towards the end of July 1651 the city of Barcelona, seeing that the Castilians were coming to besiege it, and since at the beginning of a siege the general of the enemy army usually sends a herald to the city to demand its keys, a city council meeting decided to make a vow to Our Lady of the Immaculate Conception, as the protectress and sustainer of Barcelona, to give her silver keys to the city's four main gates. At that time the only ones which were open were Saint Anthony's Gate, the Sea Gate, New Gate, and Angel Gate, while the others were walled up, except the one at the naval yards[161] which was used for the defense of Montjuich. Thus the councillors and the governor went with the available members of the Council of One Hundred, who were few because of the plague,[162] and presented her with keys made of iron while the silver ones were being made. These the lord councillors brought and presented to her on Saint Louis' Day, which is August 25. When the mass ended in the Cathedral, they sang some motets and a *Salve Regina.* They also had painted a huge portrait of all six of that year's councillors and the lord governor and the Council of One Hundred with Our Lady of the Immaculate Conception, along with an angel giving her the silver keys in the name of the city, and they also presented this picture to Our Lady.[163] May she defend and protect us in all our needs, particularly those which we had then, which were famine, plague, and war, all at the same time.

As the plague slowly ebbed in Barcelona, its inhabitants celebrated their good fortune. However, the epidemic continued to rage throughout much of the rest of Catalonia.

[61r.] Of the Second *Te Deum Laudamus* Celebrated in the Cathedral for Good Health After the Plague

You will find above the first thanks the city of Barcelona gave to Our Lord God for the favor shown us in remitting the harsh punishment of the plague in the midst of the heat. The plague subsided all over little by little until late March [1652], and the Jesus hospital was in good shape and for some time no sick persons were brought there, as no one was ill. Thanks to this, one knew that Our Lord had removed the hand of His divine justice from the city. And when the Plague Board saw this it decided to dismiss the last physicians and barber-surgeons the city had hired to take care of the sick. By that time there was no one left except one physician and a surgeon and an assistant, for as the plague subsided they had been letting the doctors and surgeons go and in the end only these three were left. And seeing that it had been a while since they had work to do, it decided to dismiss them on April 6, 1652. And on April 8, which was the Monday of the week after Easter when the city celebrated the feast of Our Lady of March, which had fallen during Holy Week that year, the councillors went to mass in the Cathedral. When it was over they had a *Te Deum Laudamus* sung and held a procession within the Cathedral and its cloister to give thanks to Our Lord and most blessed Mary for the favor shown us of completely stopping the plague. May He be pleased to maintain through His divine mercy the good health of the city and of all Christians.

And since the plague is a punishment of God who chastises not only a city but also the entire province and kingdom, there were at this time very few towns and cities in Catalonia where the plague had not spread. In some places there was more and in others less, and thus it went throughout the land. And as the city saw itself relieved by the grace of God it began to put plague guards at the Sea Gate and Angel Gate which, thanks to the siege, were the only ones open. The Council of One Hundred began as usual by watching out for people coming from places on the coast where the plague had taken deep hold, particularly Mataró,[164] where virtually no one was left. And at this time a boat came from Mataró carrying wine and some people fleeing from there, and they did not want to let them unload at the docks. Instead, they made them bring the boat up to the sea wall, where they set up a tent and burned their clothing. As right there one of them fell sick, they made him go to Jesus Street while the others had to return right away to Mataró or wherever they wished. And in this fashion they dealt with those who came from where they knew there was plague.

The plague was in the enemy army around Barcelona throughout the siege, especially in the area around Sant Martí, where they had their plague hospital. Many died and they well deserved it because although there had been so much plague in Sant Martí and Sant Andreu when they arrived, they did not let that stop them from entering the houses there to sack them, and in this way they caught the plague. I think this was the reason why the plague remained in Barcelona so long, for people were constantly coming from the enemy army into the city, either as deserters or prisoners of war or as heralds and drummers and other things that go with war. This was the reason why the plague took so long to subside in Barcelona. And the city, knowing the harm caused by these things, ordered that any soldier coming from the enemy army either as a deserter or prisoner, or one of our soldiers who had been ransomed, or any refugee was made to pass through an oven along with the clothing he brought. To this end they set up some large ovens near the royal ovens in front of the Street of the Holy Spirit. And they put some guards at the Angel Gate where everyone entered, and when people came from the enemy lines these guards would accompany them outside the wall to where the ovens were. There they found some other men who made them all undress. Then they made them pass through a warm oven while they put their clothes in a hotter one until they knew they were clean. Afterwards they got dressed again and each one went wherever he had to go. Thanks to this the situation improved greatly. They also had the streets thoroughly cleaned of their filth, and carts went back and forth picking it up and removing it, and they had all the houses so thoroughly cleaned that we can say that for once Barcelona was actually somewhat clean.

Barcelona finally surrendered to the besieging Castilian army on October 12, 1652. Shortly thereafter, the municipal authorities organized a procession to honor its patroness. Life in the city was slowly returning to normal.

❧ [92v.] Of the General Procession Held in Barcelona on the Day of Our Lady of the Immaculate Conception

With the great travails the city suffered, including war and plague and famine and other things, the Council of One Hundred decided some time ago to take as advocate and protectress of the city Our Lady of the Immaculate Conception so that she could give us shelter and help at all times and against all distresses. Thus it ordered that a general procession, like that of Corpus Christi, be held on the feast of Our Lady of the Immaculate Conception along with a great celebration. Yet ever since this was voted there had not been any opportunity to hold this procession, nor any other, thanks to the travails of the plague and war. And now the occasion arose, thanks to the surrender of the city and our having returned to obedience to Spain, which one can assume that Our Lady permitted as the best means for our preservation and peace. It was thus decided to hold the procession on the day of Our Lady of the Immaculate Conception in 1652, and this procession was held on that day with the same solemnity as on the feast of Corpus[165] except that the monstrance held an image of Our Lady of the Immaculate Conception. In all else everything was done exactly the same.

By late 1653, the plague had virtually ended among Barcelona's populace. However, it continued to afflict the army quartered within the city. Thus a new pesthouse had to be prepared.

❧ [104r.] How in Barcelona They Dismissed and Ceased to Pay the Plague Physicians and Surgeons and Cleaned the Pesthouse Located in the Monastery of Jesus and How Shortly Thereafter the Plague Once Again Flared Up in Barcelona and They Set Up a New Pesthouse in the Monastery and Church of Our Lady of Nazareth

On Saturday September 13, 1653, they dismissed all the physicians and surgeons and plague deputies hired by the city during the plague because there was not a single person sick with the plague in all Barcelona or in the Jesus pesthouse. For some time now no one had been taken there nor were there any new sick persons, just some convalescent patients now cured of the plague, and they returned well purged and fumigated. And they thoroughly cleaned and fumigated the Jesus monastery, and once it was completely clean the friars returned to live there just as they had before.

After the city of Barcelona was clean and purged of the contagion as has been said, it happened that in mid-October the soldiers (and in particular the Irish troops[166]) returned to Barcelona from the siege of Girona.[167] And since they are such dull people and so unkempt and so ill-prepared for hardships, they always carry the contagion with them.[168] They dispersed throughout the city, especially underneath the portico in the New Square and in the auction square. There were three or four hundred of them, all somewhat sick and completely lost, for nobody wanted them in their homes. They could be found in other parts of the city as well, with many women and children, and it was a pitiable sight to see. In the evening those who were well went outside the city to gather bundles of herbs, and at nighttime they slept in the New Square and other parts of the city, while during the day they walked through the city selling the herbs. Often the sick did the same, yet still they kept doing it, and in this fashion the plague spread throughout the city. And when the Irish fell sick they were taken to the general hospital. It was also said that at the same time ships came from Valencia carrying many Irish soldiers. And since they are so worthless they suffered greatly while at sea, and when they disembarked they were in poor shape, most of them being very sick. And they had to take many of them to the hospital.

Thanks to these and the ones spreading their sickness around the city who were later taken to the hospital, the plague started up again in the general hospital. For this reason, and their being so close to all the people who weren't sick, they didn't open the hospital for visits this year on its holiday, especially when they saw that the plague lingered on there. And once the pesthouse of Jesus was all clean and purged and the monks had returned, they did not want to let the convent get dirty with plague again. Since the plague kept getting worse they had to decide where to set up a new pesthouse so as to separate the plague-stricken from the other patients. They found no better place for the time being than the monastery of Our Lady of Nazareth near the Butchers' Street. This was a monastery belonging to the monks of Poblet.[169] Thus they set up the pesthouse in this monastery and brought many of the Irish there. Most of the rest fell ill from the bad food they ate, for they ate nothing except hardtack and the herbs they picked from local gardens. Since they still slept in the streets, when the city saw that the plague kept spreading it had no choice but to petition His Highness[170] to remove the Irish soldiers from the city. This was done, and they were sent to quarters assigned them in Sants and Hospitalet, and afterwards every place they had been was cleaned.

The plague did not stop spreading throughout the city, particularly among the soldiers, and it spread to some houses as it grew in virulence. By mid-November there were enough sick in Barcelona for the city to have extreme unction for the plague sent by itself, without any canopy, just the priest and his assistant. And just as before they kept the Hosts for the plague-stricken in the chapel of Saint Agatha.[171] Afterwards they moved them to a house on Saint Paul's Street, from whence they were taken to the sick throughout the city. They set up the pesthouse in the aforesaid monastery of Nazareth, and hired plague doctors and surgeons and orderlies. They put two Capuchins[172] to serve as overseers and vicars to administer the sacraments, and they held this office while the plague lasted. They did much to ensure the proper functioning of the pesthouse, so that no one would talk about it the way they did about the poor administration at the pesthouse of Jesus. Thus in regard to the care of the sick and to the respectful and good behavior of the orderlies, at no time did they commit a single bad act, great or small, like they did in Jesus. Instead, they did everything they had to do with great charity and to the great satisfaction of all the patients.

The plague kept spreading, so much so that everyone thought that it would get as bad as the last time, but thanks to the [104v.] good order and care of the district officials[173] upon whom was conferred full control over the pesthouse, it was the will of Our Lord that it not spread too much. During the six months while the pesthouse of Nazareth was open they never took any sick or dead person there during the day, only at night. Their policy was to alert the plague wardens when people who fell sick from the plague, whether soldiers or civilians, could not take care of themselves at home. Later at nighttime they went with the grave diggers and they carried them to the pesthouse. They also took along a cart and removed all the mattresses, blankets, and sheets of the sick and took them to Nazareth. They also nailed shut the sick person's house and did not allow anyone who had been in contact with the sick person to leave the house; instead, they made them undergo quarantine right there. They also burned the clothes that the patient wore when he got sick in front of his house at nighttime, and if anyone died of the plague they took the body at night to be buried in the graveyard of Nazareth along with the mattresses and sheets. The following night they would come to burn the wooden bed frame and curtains and the clothing and everything the sick person might have touched, regardless whether they were Spaniards or Catalans. They made no exceptions for anyone. A councillor was present whenever they burned the goods of Catalans, while an army officer was present whenever they burned the goods of Spaniards, and they all went around together at nighttime, with the one watching over the other. And they burned many goods of great value. And they made the people in the houses of the dead or sick undergo quarantine in the very same houses until the councillors or whoever was in charge decided that they had been there long enough. Afterwards they made them fumigate and clean the houses along with those who had been shut up inside, who had to change and burn the clothes they were wearing. All this was done very correctly, and in this way they undercut the plague.[174]

The plague increased greatly during the full moons of November and December, so much so that the sick could not fit into Nazareth. There were some four or five hundred of them, and it was feared that the plague would worsen. They were supposed to hold a great celebration and procession in the Cathedral on the feast of the Immaculate Conception to fulfil the vow the city had made in 1651, and since for lack of opportunity they had never done so they wanted to do it this year, 1653. But with the plague spreading so much they couldn't do it and thus had to leave it for another time. They finally did it on February 23, 1654, as will be explained in the proper place below. And it will also be explained below that the plague stopped and a procession was held on May 2, 1654.

Barcelona continued to celebrate votive processions in gratitude for protection against its recent sufferings and in anticipation of future ills.

1654
[106v.] How Barcelona Celebrated Our Lady of the Immaculate Conception as the Protectress of the City

Some time ago the city of Barcelona had taken the Virgin Mary of the Immaculate Conception for its protectress not only against the plague but also against war and famine. As one can see above, the city had promised during the time of contagion and the siege of Barcelona to hold a great celebration in the Cathedral on her feast day. Yet the city had never held this celebration, thanks either to the persistence of the plague or the great travails of the war. Seeing that the contagion was not subsiding, for there were many sick in the pesthouse of Nazareth and throughout the city and there was no stop to it and many were dying, the city realized that the plague would not end were the celebration not held,[175] so the Council of One Hundred thus decided to celebrate the festivity as best as it could and set aside a day for it. This was Monday, February 23, 1654, which was not her feast day, but the plague still kept raging as was noted above, albeit at that time it was beginning to calm down a bit. Thus it was held on that day, although not with the pomp and expenditure that would have been undertaken if the city had not had the many expenses that it did. Instead, it was done rather plainly, like what was usually done on the feast day of the Immaculate Conception. Thus on February 23 a solemn mass with sermon was held in the Cathedral, with many palm branches and candles, and was attended by the Most Serene Highness Don Juan of Austria and the lord councillors. And the cloister and the altar of Our Lady were adorned as if it was her own feast day. After the mass was over, a solemn procession was held along the route followed every year for Our Lady of the Immaculate Conception, to Saint James's Square and then to the Booksellers' Street and the King's Square and in front of the Inquisition[176] and returning to the Cathedral, with the monstrance containing the ornate image of Our Lady, and His Highness following the canopy with a small torch in his hand. The celebration was held to the great delight of all. May the Virgin Mary intercede for the favor of her Son, that He provide remedy for the plague and be pleased to bestow peace and quiet upon this land and on all of Christendom.

The plague slowly subsided and each day it kept improving, and within a few days the pesthouse of Nazareth was closed and cleaned.

[107v.] How a Solemn Procession Was Held in Barcelona in Thanksgiving for the End of the Plague and the Closing of the Pesthouse of Nazareth

[108r.] Our Lord extended to us the great favor of ending the contagion in the city of Barcelona and particularly in the hospital or pesthouse of Our Lady of Nazareth, for truly it seemed that since the plague had so taken over the city and pesthouse during the winter, that in the summer it would rage again as it had in the year 1651. It was the will of Our Lord, thanks to the intercession of most Holy Mary and some pious persons, for the plague to stop so completely that not a single sick person remained in the pesthouse, nor was anyone known to be sick with plague in the entire city, and all its houses were clean and purged. Thereafter the pesthouse and other buildings at Nazareth were ordered to be cleaned and purged and many mattresses and clothes were burned. Those which could still be used were cleaned and purged. Once this was done the Capuchin fathers who were in charge of the pesthouse and the district officials who oversaw its administration and expenses wished to hold a solemn celebration and procession before it closed to give thanks for the great favor Our Lord had shown us. Thus the Council of One Hundred ordered a great celebration, which was held to the immense joy of all in the following fashion.

The plan they had for the procession was to return the Holy Sacrament, which the fathers had kept in a house or chapel on Nazareth Street to give to the sick in Communion. Thus they wished to hold a solemn procession to carry the Holy Sacrament in its monstrance and to display it in the Cathedral. The day fixed for this procession was Saturday, May 2, 1654. As the day before was the Day of the Apostles[177] and the day after was Sunday, they decided to hold three celebrations, and issued orders several days in advance about how the main procession was to be held. It was to be like Corpus Christi, and everyone was asked to light up their houses at nighttime beginning on Friday night and to continue this for the next three nights, which everyone did quite willingly and with great happiness. And on the Day of the Apostles they set up a richly decorated altar in front of Nazareth Street next to the episcopal seminary, which closed off the street. Afterwards this small square was draped with veils. The altar was decorated with the ornate livery of the *Diputació*, and since the street was in the parish of Saint Mary's,[178] the priests from that church decorated the altar very lavishly with a great deal of silver. The afternoon of the next day, which was the second of May, the priests from Saint Mary's went in solemn procession to say mass at the altar, going through the streets where the main procession would later pass with the giants, dragons, serpents, devils, and horses[179] and everything else just like on Corpus Christi Day. And at night everyone lit up their houses with many candles (even though this was a time when there was little wealth and much hardship in Barcelona, everyone made an effort to do what they could). That night Our Lord was pleased to water the land with a beautiful and abundant rain. And in the morning the weather got better, and since they had had to dismantle the altar because of the rain, they set it up again. After decorating it and fixing it up so mass could be said, Father Chrysostom of Barcelona, the Capuchin monk who had held the office of vicar in the pesthouse, said mass at the altar and consecrated the Holy Sacrament, which was to be displayed in the monstrance there. And after the mass was finished he placed the Holy Sacrament in the monstrance and left it on the altar amid the bright light from four torches and four large candelabra.

When the altar was ready, the priests from Saint Mary's parish came in solemn procession, carrying their banners before them along with the cross and all the clergy and vestrymen,[180] and they said mass there accompanied by an organ and two choirs. The preacher was Father Ignatius de Sant Feliu, a Capuchin monk and a most celebrated preacher,[181] and he spoke of the way in which we should give thanks to Our Lady for the favor she had recently shown us in halting the plague. His sermon pleased everyone greatly. Once the mass was over, the same procession of parish clergy and officials returned with great devotion.

That same morning a great festivity was held in the Cathedral, where a solemn mass was celebrated like that of Corpus Day, with the attendance of His Highness and his noble retainers along with the councillors and leading citizens. The sermon was delivered by the great Jesuit, Father Cabrera, the most celebrated and applauded preacher in Barcelona at that time, who also spoke of the great thanks we should give Our Lord for the many manifest favors He had shown us when He stopped the plague that had been raging so much, and that we should continue to render thanks so that Our Lord would continue His favors.

That afternoon the entire procession assembled in the Cathedral just as it does on Corpus Christi Day, that is, beginning with the parish priests along with the regular clergy and all the confraternities with their banners or flags. The procession started to leave the Cathedral around four o'clock, with everyone in his customary place, marching in front of the bishop's palace and [108v.] through New Square and through Saint Anne's Square and in front of Mount Zion[182] and Saint Anne's Street and in front of the university and down the Butchers' Street until it reached a street which empties onto Nazareth Street, where the altar was set up. The procession then passed in front of the altar, where it turned to the left down a street which passes in front of the convent of the Angels.[183] The entire procession passed in order before the altar. And when it was time they stopped and took up the Holy Sacrament and placed it in the monstrance beneath the canopy carried by the lord councillors, just like on Corpus Day. Afterwards the procession started up again and headed down the street in front of the Angels convent, and when the Host reached that point, His Highness came out of a house in front of the convent where one of his lieutenant generals lived, and his pages came out with him with their torches and walked in front of the Host, and His Highness walked behind it carrying a small torch in his hand, accompanied by all his noble retainers and guards. The procession continued in front of the Angels convent and down the street which empties into Carmen Street at the corner where the Monastery of Carmen stands.[184] It then crossed the patio of the general hospital and went down Hospital Street and crossed the *Rambla* and continued through the *Bocaria*[185] and *call*[186] and in front of Saint James's and down Booksellers' Street and through the King's Square in front of the royal palace and in front of the Inquisition and entered the Cathedral through the main door, where with much singing and joy on the part of everyone the procession ended. The Holy Sacrament was put away, and His Highness returned to his palace[187] while everyone else went home.

The illuminations continued that night and the following one to the great joy and content of all. May God favor us by continuing to protect us and all Christians from any other such plague.

❧ Of Another Procession of Thanksgiving for a Vow Made to Saint Nicholas of Tolentino

The members of the confraternity of glorious Saint Nicholas of Tolentino,[188] located in the monastery of Saint Augustine[189] in this city, saw the great miracles that Our Lord had worked during the plague in Barcelona through the intercession of the aforesaid saint, as one can read above under the year 1651. The members of this confraternity therefore decided during a meeting held in that church that if Our Lord were pleased to deliver the city from the evil of contagion, that as soon as the Cathedral held a procession in thanksgiving, then they would promise to hold another, most solemn one, and would carry the huge statue of Saint Nicholas which crowns the altar in his chapel in the church of Saint Augustine.

The members and officers of the confraternity saw that the Cathedral held its procession as reported on this same page. Since the Cathedral did so on May 2, they wished to hold theirs on Sunday the tenth of the same month in 1654. And when the procession had assembled at the church of Saint Augustine and was ready to start, it was the will of Our Lord that a huge and furious rain fall, and it could not take place that afternoon and had to be postponed until the following Thursday, which was Ascension Day, May 14, 1654. So on that day the procession assembled again in the church of Saint Augustine, and included all the banners of the confraternities in that church along with the large statue of Saint Nicholas under a canopy carried by the councillors. There was a huge crowd of people and much bright light from the confraternity members and other devout persons, who carried over three hundred torches among them, and it was a great thing to see so many people along with the light. The procession left Saint Augustine going up the Bòria[190] and the King's Square, and then it entered the Cathedral. After thanks were given to Our Lord, it left the Cathedral and passed in front of the bishop's palace and Saint James's Square and Regomir[191] and Broad Street and entered Saint Mary's,[192] leaving it through the door in front of the parish graveyard and through the Born and Llansana Street and the Curriers' Square and back to Saint Augustine where, to the great joy and content of all, they returned the statue of the saint to its altar. May God be pleased through His intercession to deliver us and all Christian peoples from such plagues.

APPENDIX I

Other Voices: An Anthology of Popular Plague Texts

The following are excerpts from diaries and autobiographical remembrances of plague written by artisans and other members of the popular classes in late medieval and early modern Europe. Although none of them contains as detailed an account of plague as that written by Parets, they nevertheless provide interesting points of comparison with the tanner's narrative.

THOMAS PLATTER[1]
Ropemaker/Printer
Zürich-Porrentruy, Switzerland
Mid-Sixteenth Century

When I returned home to my wife, she was happy to see me, for the parish priest had fallen ill with the plague. . . . As for myself, I had already gone through the same thing several years before. When I was a schoolboy in Zürich, there was such a terrible plague that nine hundred bodies were buried in a single grave next to the *Grossmünster*,[2] and seven hundred in another one. I thus returned home along with some others from the country. I found a boil on my leg, and thought that I too had the plague, and they would hardly let me enter the house. I then went to my aunt Fransy's house in Granges, and fell asleep eighteen times during the short walk from Calpentran[3] (a small village at the foot of the mountain) to Granges. My aunt made me a dressing of cabbage leaves, and with God's help I was cured and no one else fell ill. But neither I nor my aunt were allowed to go anywhere for six weeks. I was also in Zürich another time when there was an epidemic of plague, when I was staying at the house of the mother of Rudolphus Gwalterus.[4] And as she didn't have many beds, I had to sleep with two young girls who both caught the plague and died right next to me, although nothing happened to me. . . .

After we had been there [Porrentruy][5] some twelve weeks, and our baby had begun to take his first steps, the plague took him, and he died on the third day of his illness. He had many convulsions, and we saw how cruelly he suffered. When he died, both my wife and I cried from sorrow and also from joy that he had been delivered from his sufferings. His mother made him a pretty little garland of flowers and the schoolmaster of Porrentruy buried him behind Saint Michael's. Since both of us remained sad and my wife was no longer happy the way she used to be and no longer felt like singing, my master [a doctor][6] said, "Your wife is no longer happy and my wife is afraid, fearing that she is so sad that she will wind up also catching the plague (which was then raging in Porrentruy). I advise you to take her elsewhere." That is what I did; I took her to Zürich. We didn't spend more than five *batzen*[7] on the trip, but I returned to Porrentruy all the same. I arrived at my master's house on a Sunday evening. He was sitting alone at the table; he was full of wine and said, "Oh, Thomas, you were wrong to take Anna so far away" (although it was he who ordered me to do so). "As soon as she was gone my wife caught the plague. She is in bed upstairs and has a large plague boil on her leg." My master was very afraid and that is why he got drunk every day, so he wouldn't have to think about it. He was drunk most of the time; when we ate out he would drink a great deal, and then he would order the waiter to take him down into the wine cellar where he could drink even more. When afterwards we returned home, he would send for even more wine (he had none left in his wine cellar) and would sit in his garden until after midnight with only his shirt on and there he would continue to drink.

On Monday night (I had returned on Sunday) he caught the plague, and told me, "We are leaving for the country." When we left the city gate behind, he announced, "We are going to Delémont." For that is where the bishop had gone to flee the plague. That day we reached the nearest village on the way to Delémont,[8] a mile or so from Porrentruy. We spent the night there. He did not wish to eat anything, for he was quite ill. He had not told his wife that he was going to leave, and I didn't realize what was going on until we left the city behind us. The next day, we rented a horse and, when crossing the mountain between Porrentruy and Delémont, he fell from it, as he was a big man, heavy and very sick. When we reached the village closest to Delémont he sent the horse back and walked on foot the rest of the way. At the city gate they didn't want to let him enter until he sent word to the bishop's palace that he was there. After the bishop ordered them to let us enter, we went to his palace. He welcomed my master, who sat next to the bishop during the evening meal, although he ate very little. The bishop then asked him, "Sir Doctor, do you feel well? You are not as cheerful as in the past." He replied, "I had a fever yesterday during my trip, and I drank a bit; that has made me sick." When he left to go to bed, the bishop asked him if he would like to go hunting the next day. The doctor replied, "Yes, my Lord, as I hope to feel better." We were then taken to a large bedroom, where my master slept in one bed while I slept in another. He was very sick during the night and vomited in his bed. They had placed two large drinking glasses on the table, one full of wine and the other with water. In the morning, the doctor could barely get up; I washed the bedding as well as I could with the wine and water so they wouldn't see what happened.

The bishop went hunting on horseback and came back early. After he climbed down from his horse he called me and asked, "Tell me, Thomas, did you lose your baby in Porrentruy and is the doctor's wife sick with the plague?" (He had learned this while hunting.) I said, "Yes, sir," "Why did the doctor come here? Tell me, does he also have the plague?" I said, "I do not know, he has not told me." "Just look what you have done," he said, "go to your master right away and take him far away from my house!" I then walked all around the little town; no one was willing to take him in, and everyone asked me about the disease my master had. There was a woman who kept an inn (I think it was named The White Cross) who told me to bring him there; she put him to bed and treated him with the respect he was due. My master then told me, "Thomas, go to my wife, and tell her that if she wishes to see me before I die, she had better come quickly."

When I reached his house in Porrentruy and told her this, she became very angry. "The scoundrel," she said, "he is like all these bumpkins.[9] When I was in trouble, he fled from me. I cannot see him and I have no desire to do so. May he get what God has in store for him!" I said, "Madame, I think that he is going to die. You have many debts here and in Basel; they will take everything you have. Give me what you have, I will take it to Basel and keep it for you there if he dies." She then gave me my master's case book, to which he attached great value, and three shirts; she had made them by hand and they were quite beautiful. She also gave me a silver spoon, some handkerchiefs, and I don't know what else. The book was the most valuable thing as far as I was concerned, and I had already planned to copy it.

[Shortly thereafter Thomas returned to Basel, after visiting his master for the last time.] Upon my arrival there I learned that he died the very day I left him. He was buried in Moutier[10] with the honors due him as a doctor. God had deprived this man of all earthly help, as he had neither a surgeon nor any drugs with him, and he had left many drugs behind in Porrentruy, where he had always kept his own personal pharmacy.

GIOVAN AMBROSIO DE' COZZI[11]

Popolano [Artisan]
Milan, Italy
1576

In the name of Our Lord, on the twelfth day of July, in the year 1576. Giovan Ambrosio de' Cozzi, the son of the deceased Pietro de' Cozzi of the quarter of the *Ortolani* [gardeners] near the Como Gate in Milan, etc. I write this in order to remember the year the plague was in Milan, 1576. . . .

I, Giovan Ambrosio Cozzi, write the following memoir of the members of my family who died [during the plague]. The first was my aunt Franceschina de' Cozzi and then her little daughter Angela and then Pietro Antonio and Battista her sons. Not a single child escaped death in that household.

Four members of the household of my uncle Baldassare died. They included Baldassare and his father,[12] Maddalena and Dionigi and his son Giuseppe, and thereafter a sister-in-law of his with one of her children who was named Giacomo Antonio de' Cozzi.

In the household of Gianpaolo de' Cozzi, his wife and three of their children died. She was named Isabella, and the children were named Gianpietro, Margarita, and Angela. These, along with a cousin named Ventura, were all members of the Cozzi family who died of plague.

Some fifteen members of the Cozzi family died of plague; may God not will such a thing ever to come again. Be warned that it is said that plague strikes Milan every fifty years. . . . If you happen to be there, or in any other place when plague first strikes, I advise you to flee for as long as you can, in order to save yourself.

Among the various uncles and cousins of my wife's family, some sixteen died; between the Cozzi family and my in-laws, some thirty-one of my relations, both male and female, died.

GIAMBATTISTA CASALE[13]
Carpenter
Milan, Italy
1576

The plague suddenly began in the quarter of the *Ortolani.* And everyone said that it was thanks to certain renegade Christians who came from Turkey and brought the disease to many cities and other places. First they brought it to Melegnano,[14] and from there to Milan. And it began in the aforesaid quarter of the *Ortolani*, and took such hold there that people dropped like flies.[15] It was so bad there that the rulers of Milan were forced to try to keep the disease from entering the city by closing off the entire area outside the Como Gate. Thus one night around five o'clock the Marquis of Ayamonte (who was the governor of Milan) went with all his foot soldiers and cavalry in great number to close off that quarter. The residents there fiercely resisted being shut up, and large numbers of soldiers were ordered to block the streets, while the death penalty was decreed for anyone who left the area. Then they began to open [the monastery of] Santo Gregorio[16] as a pesthouse. And it is true that the first person to go there was the innkeeper from Melegnano [who had brought the plague to Milan]. But it was the will of God to send this flail to punish us, and justly so, thanks to our enormous and most evil sins. Although we had been admonished during so many years, we never did anything to correct them. Thus God allowed the plague to enter the city, and go down the street from the Como Gate to the Orefici,[17] and from there it spread throughout the entire city, so much so that one heard nothing else in Milan save the cry "the grave diggers are coming with their cart," which carried the sick and the dead to the pesthouse. . . .

The viceroy and the most Excellent Senate, seeing that the disease kept spreading, ordered all schools[18] closed. Still, the sickness persisted, so they issued another order that all women and all children under fifteen years of age should not leave their houses for any reason whatsoever. Thus beginning on October 1 no women or children left their houses, except the midwives and others who had license to move about, but not even this stopped the plague. . . . With all this the true marvel and the most wondrous thing was to see the most Illustrious and Reverend Cardinal Borromeo, Archbishop of Milan, order the saying of prayers around the clock for all Milan and its diocese. He did this to see if through the means of holy prayer the anger of God might be placated, and through

His endless mercy He might put an end to the plague. Thus the Archbishop ordered everyone to pray. . . .

The most Illustrious Cardinal ordered all these holy processions and prayers to be carried out to implore the grace of God, that He be pleased to have mercy on His people and free them from so great a punishment as the plague. One heard or saw nothing else in Milan from morning to night save the plague, and the dead and plague-stricken being carried to Santo Gregorio or to the huts built outside the city, and the closing up of houses by nailing their doors shut and forcing their residents to undergo quarantine. One heard nothing else in Milan save talk of the grave diggers, whose only chore was to transport the plague-stricken and the dead to these places.

I remember how on October 9, 1576, which was the day on which a general procession was held to worship the most Holy Nail[19] of Our Lord Jesus Christ . . . I, Giambattista Casale, walked along in this procession out of devotion and was present for the holy blessing given at its end by the most Illustrious Cardinal Borromeo,[20] and the procession went on until ten o'clock at night before the most Illustrious Cardinal gave the benediction. I was going home for dinner when a brother from the confraternity named *messer* Giovanni Pietro de la Maldura told me that we should meet at the Cathedral after the procession ended. He gave me some bad news, which was "that the officers of the Plague Board had gone to visit some sick persons in your home, and they have concluded that the disease is plague, and they quickly ordered the house and shop closed, and they have been asking where you are, and I told them that you were in the procession, so they will be waiting for you to return home, and you will be shut up with everyone else to undergo quarantine." And I told him, "First, I want to finish my prayers, and to see the end of such a devout and holy procession and to be blessed by the hand of the most Illustrious Cardinal with the Most Holy Nail of Our Lord, and then I will obey these officers and shut myself up in my house in order to undergo quarantine. And I hope through the Lord Jesus Christ that thanks to the holy blessing of His most Holy Nail that I myself and all my family might escape such a great punishment through the mercy of God. May all this be done for His greater glory, and for the salvation of my soul and those of my family." Thus we came to be shut up for about eight days after the grave diggers came to take away the clothing of the sick. . . .

The grave diggers left behind a small fire with herbs to fumigate the beds where the sick had died, and around midnight the room caught on fire, and we woke up to find ourselves trapped upstairs in our room by the fire. None of our neighbors wanted to help us out of fear of the disease, and we were there in our room frightened to death as the fire worsened. Seeing that our only source of help was almighty God, we threw ourselves at the feet of His boundless mercy and invoked the glorious Virgin Mary and all the saints who would design to intercede for us with His divine majesty. In this manner we were saved and barely escaped such a great danger. Armed with our belief in His divine goodness, we all made the sign of the Holy Cross, and while some of us stayed there to pray, I and *messer* Giovanni Dominico Tura, the owner of the house, and *messer* Christoforo Della Porta, all of whom had been cut off in our rooms with our families (that is, with our wives, children, and mothers), went around to check all the rooms, and we found a fire burning in the room just opposite mine and in the room right above, where the fire the fumigators lit had begun. We then began to take heart and break and smash everything in the path of the fire, and we also began to cry out to our neighbors to have the officers of the Plague Board send some of the grave diggers, that is, the only persons who were allowed to have contact with the dead and the sick. As soon as the officers heard what had happened they sent some soldiers along with three grave diggers, and together they came to our building. . . .

Our salvation can be attributed only to Almighty God, who was pleased to help us thanks to our having an *Agnus Dei* blessed by His Holiness the Pope. I had placed it where there was the greatest danger of fire, and thanks to its great power the fire suddenly began to die out. . . . Thus Our Lord God, in His great goodness and mercy, extended to us, myself, and my wife Catelina and our three children Zanevera,[21] David, and Angela, the grace of allowing us to survive the quarantine safe and sound. *Soli Deo honor et gloria.* This quarantine began on October 9 as I said earlier, and ended on November 17 of the same year.[22]

ANGELO MICHELE RISI[23]
Glassmaker
Bologna, Italy
1630

Today, the eighth of August, my dear and beloved daughter Elena Teresa died [of plague], ten days before her thirteenth birthday. She had the good sense of a woman fifty years old, and was in all respects virtuous and most lovable thanks to her most rare gifts. May God make me worthy of her holy prayers. . . .

Today, the eighteenth of August 1630, the youngest of my daughters, Anna Maria, died at the age of eight years, ten months, and eight days . . . hers was a quick death, and at the end she raised to heaven eyes so bright and beautiful that she appeared not sick but well, and I firmly believe that, seeing an angel or something else in Paradise, she too flew up there to sing the praises of her Creator. . . .

Today, the twentieth of the same month, at eight o'clock in the evening my dear companion and wife Lucia Conventi died at age forty-one. May God rest her soul. I can write no more. . . .

[Risi himself died five days later, two days after the death of his eighteen-year-old son Ercole.]

JUAN SERRANO DE VARGAS[24]
Printer
Málaga, Spain
1649

I will now proceed to tell of the wretched contagion [the city of Málaga] suffered last year, 1649, caused (it is said) by the numerous levies of troops which passed through the port to join the royal armies. These soldiers (constantly exposed to the hardships of long journeys, along with harsh weather and bad food) arrived in the city already infected by terrible and contagious diseases. They were joined by a goodly number of other sick persons, also afflicted by lack of provisions and poverty, and most of them foreigners. . . .

To remedy this a pesthouse was set up, and the noble councilmen[25] of the city cast lots to see who would govern it and oversee the care of the sick. . . . All these gentlemen scrupulously carried out their duties day and night, without rest, not even taking time to eat. . . . One would have to write a whole book about each of them in order to consider and praise their charity and Christian zeal; my talent and abilities are insufficient, and this treatise too short, to carry this out. . . .

At this time everything in the city was confusion. The priests carried the Holy Sacrament hidden from sight,[26] while working day and night without stopping. Meanwhile, trade with the outside world and local business came to a halt, and many inhabitants fled to the countryside. . . . The plague spread, the numbers of dead and stricken increased, much clothing was burned in the city and the pesthouse, and there was no more room to shelter the sick, as more than 2,600 beds were full, according to official reports. Thus a new pesthouse was set up . . . a good number of monks went there of their own free will to tend the sick, along with many young noblemen from the city. . . . Opinions differ as to the number of dead in the city, pesthouse, and the nearby countryside. I have heard many persons claim that more than forty thousand died. . . .

I will mention a few of the endless number of pitiable cases in the city. In my neighborhood, five children and two servants died in the house of one of my neighbors, and he wound up dying in the pesthouse, where his mother (more than sixty years old) had come to take care of him. His wife, who had nursed their children, still lives. The wife of another of my neighbors died while nursing her baby, and one of their daughters also died while feeding the child, and two other children died in the arms of the Africans[27] who were carrying them to the pesthouse, and then another child along with the father, and shortly thereafter the baby himself, whom a charitable woman had taken from the home to look after. . . . In another house the father, after his entire family died, went to church, confessed and took Communion, returned home and locked the door, and then lay down and was later found dead. Many deaths of this sort occurred, because when people refused to go to the pesthouse, others didn't know they were sick, and thus they perished for lack of care.

ANDRÉS DE LA VEGA[28]
Shopkeeper
Seville, Spain
1649

I shall now speak of what caused so many deaths in so short a period of time, many more than has been witnessed in any other epidemic. I am convinced that the Devil was responsible, for when people died of plague south of here, their clothing was thrown into the river along with goods of great value, which boatmen and persons without scruples brought into our city, to Triana, the Street of the Virgins, and *Hierro Viejo*.[29] And even before this, some of these goods were taken to a house near the Golden Tower,[30] where many people lived.[31] . . . Some six days later not a single person remained alive there. . . . These were the first plague victims in Seville, and the local law officers ordered the building nailed shut. For some time afterwards nothing was heard of this incident, until (as I have said) the infected clothing began to enter the quarters mentioned above, and along with it a punishment from heaven. . . . [Soon thereafter] no one could take a step without walking on the clothing of those stricken with plague, from which one could conclude that an endless number of persons had fallen sick, and that since this disease was caused by the stars, or corrupt air, there was little you could do. . . .[32]

Among our misfortunes, I think the greatest was that the sun didn't shine once while there was plague, and if it came out it was pale and yellow, or else much too red, which caused great fear rather than consolation. It rained so much and so continuously, and everyone was so discouraged, that one began not to see anyone in the plazas and streets. . . . Everything was permitted, even though it frightens one just to say it. There was no end to the selfishness of everyone against everyone else, as a lack of charity unleashed our evil nature. . . . The shamelessness was such that it is impossible to tell of it. . . .

As I write this, today the tenth of September, the government ordered a man and a woman whipped for having stolen plague-infested clothing. . . .

When things were in this state, a great disturbance was caused by the ecclesiastical judge and his officials, for the crowd of common persons waiting to be married was so great (in forty days more than 1,500 persons got married) that the judge had to leave for a few days. Many persons, especially of the better sort, decided to wait, even though during the worst part of the plague many persons married women whom they never would have married otherwise, as being beneath their status.

I should say that this sickness respected neither sex nor age, nor high or low estate; instead, it started with those yet to be born. Most people believe that some 60,000 women died, and although most were unmarried, others were pregnant, and both mother and child died, from which I conclude (according to reliable sources) that more than 20,000 children died before being born. More than 140,000 men died. . . .

I will mention another great misfortune, which was to see many small children in the streets whose parents had died, and lacking help they went about looking for some food, and if anyone gave them something it was by throwing it to them like dogs. They could be found near the city gates, where they died either from hunger or disease. . . .

There were two reasons why I began to write these poorly shaped sentences. First, out of idleness, and second, to remember the plague that struck fifty years ago, about which I have seen nothing written, neither in recent histories of those times nor in manuscripts, nor anywhere else, nor do I know the reason why so many learned and curious men have not written about it. And although I know for sure that this epidemic will not pass in such great silence, for my own curiosity I have nevertheless wished to write about it as best as I can, even if this is such a short relation of it. . . .

[De la Vega then offers a lengthy account of the various religious processions carried out in the city during the plague.] Much of what is written here about what happened during the year 1649 (which was the year in which there was a great plague in the city) has been taken from the book written by Don Bernardo Luis de Castro Palacio, priest and head sacristan of the Cathedral of Seville, wherein he has noted the ceremonies carried out there for future reference. . . . I have also noted what I myself saw, being present in many of these processions, because during this time I was in Seville and saw all these things.

JOSÉ ESTICHE[33]
Surgeon
Saragossa, Spain
1652

The past few years have been most unfortunate and full of calamities in various countries of Europe, as some and then others have burned with the never-ending fire of war and plague. It was the will of Our Lord God justly to punish (or mercifully to visit) with the flames of both the imperial city of Saragossa, which, having suffered for twelve years the calamities caused by the revolt of its neighbor Catalonia, also suffered plague in 1652. It began in early March, and spreading with great fury among all estates, devastated the city until November, when it began to decline. Its having tempered so quickly, and its not having claimed a greater number of victims during this time can be attributed to (after God, the most holy Virgin of Pilar our patroness,[34] and the saints who interceded for us) the great zeal, care, piety, and generosity of those who, in such a stormy year, and thanks to the special providence of heaven, governed this commonwealth. These men, in a rare example of paternal care, preferring the public good to their own comforts, resolved in their council not to spare any effort, no matter how great, nor any expense, no matter how large, in their struggle against the invasion of so fierce an enemy. . . .

Thanks to the summer heat, the fire of contagion spread quickly, and so many fell sick each day that they no longer fit in the old pesthouse. . . . The city thus decided to set up another pesthouse in the largest convent, that of the Capuchins. . . . At the suggestion of the city magistrates, Dr. Francisco Huguet was hired as its physician, and I went there as the surgeon with all my household. . . . Dr. Huguet died on August 23, while on the 16th of that month I fell so gravely ill with the disease that all my friends gave me up for dead. However, God was pleased to restore my health within a few days, although the plague wreaked no little havoc in my household, taking from me my wife, a journeyman, and three servant girls. One of my brothers also fell sick, and although he almost lost his life, he managed to escape. I worked in the pesthouse without ceasing from August 2 until November 11, at all times taking care of the sick there.

APPENDIX II

A Select List of Autobiographical Plague Accounts[1]

Name	*Place/Date*	*Profession*
Thucydides	Athens, 530–527 BC	soldier/historian
Procopius	Constantinople, 542 AD	soldier/historian
Giovanni Boccaccio	Florence/Naples, 1348	writer
Matteo Villani	Florence, 1348	merchant
Guy de Chauliac	Marseilles, 1348	surgeon
Raymond Chalin de Vinario	Avignon, 1348	physician
Henry Knighton	Leicester, 1348–1349	cleric
Giovanni Chellini da San Miniato	Florence, 1437	physician/humanist
Luca Landucci	Florence, 1478, 1495, 1497–1498	apothecary
Thomas Platter	Bern, Zürich, Basel mid-sixteenth century	ropemaker/ schoolteacher/printer
Edward Hall	London, 1517–1518	lawyer
Bartolomeo Masi	Florence, 1523	coppersmith
Nicolas Versoris	Paris, 1523	lawyer
Benvenuto Cellini	Rome, ca. 1525	goldsmith
Marino Sanudo	Venice, 1528	noble
Hermann von Weinsberg	Cologne, 1541	lawyer
Wolfgang Vincentz	Breslau, 1542	silversmith
Andrés Laguna	Metz, 1542	physician
Pieter van Foreest	Delft, 1557	physician
Claude Haton	Provins [France], 1561, 1582	noble
Edmund Grindal	London, 1563–1564	cleric
Laurent Joubert	Southern France, 1564	physician

Name	*Place/Date*	*Profession*
Pero Roiz Soares	Lisbon, 1569	writer
Giovanni Filippo Ingrassia	Palermo, 1575	physician
Filippo Giacomo Besta	Milan, 1576	unknown
Giovan Ambrosio de' Cozzi	Milan, 1576	unknown *popolano*
Giambattista Casale	Milan, 1576	carpenter
Rocco Benedetti	Venice, 1576	notary
Pierre de l'Étoile	Paris, 1580	judge
Michel de Montaigne	Bordeaux, 1585	writer
Pere Gil	Barcelona, 1589	cleric
Simon Forman	London, 1592	astrologer
Jean Patte	Amiens, 1596	tax collector
Rodrigo de Castro	Hamburg, 1596	physician
Jacopo Strazzolini	Cividale [Italy], 1598	cleric
Andrés de Cañas	Burgos, 1599	councilman
Fabio Nelli	Valladolid, 1599	banker
Martín González de Cellorigo	Valladolid, 1599	writer
Antonio Ponce de Santa Cruz	Valladolid, 1599	physician
Martín de Senosiain	Pamplona, 1599	municipal secretary
Agustín de Rojas	Seville, 1599	playwright
Thomas Dekker	London, 1625	playwright
René Gendry	Angers [France], 1626	surgeon
Alessandro Tadino	Milan, 1629–1632	physician
Filippo Visconti	Milan, 1630	noble
Giuseppe Ripamonti	Milan, 1630	cleric
Federico Borromeo	Milan, 1630	cleric
Agostino Lampugnano	Milan, 1630	cleric
Luca di Giovanni di Luca Targioni	Florence, 1630	merchant
Francesco Rondinelli	Florence, 1630–1633	librarian
Giovanni Angelo Michele Risi	Bologna, 1630	glassmaker
Pietro Moratti	Bologna, 1630	cleric
Giuliano Ceci	Pescia, 1631	notary (?)
Cristóbal Gómez de la Hoz	Málaga, 1637	scribe
Henri Campion	Franche-Comté and Auvergne [France], 1637	soldier

Name	*Place/Date*	*Profession*
Robert Lenthall	High Wycombe [England], 1647	cleric
Francisco Gavaldá	Valencia, 1647–1648	cleric
Andrés de la Vega	Seville, 1649	shopkeeper
Gaspar Caldera de Heredia	Seville, 1649	physician
Juan Serrano de Vargas	Málaga, 1649	printer
Alonso de Burgos	Córdoba, 1649	physician
Jeroni de Real	Girona [Spain], 1650	noble
Miquel Parets	Barcelona, 1651	tanner
José Estiche	Saragossa, 1652	surgeon
Girolamo Gatta	Naples, 1656	physician
Andrea Rubino	Naples, 1656	physician
San Gregorio Barbarigo	Rome, 1656–1657	cleric
Girolamo Gigli	Rome, 1656–1657	patrician/writer
Marco Gastaldi	Rome, 1656–1657	cleric
Nicolo Spinola	Genoa, 1656–1657	merchant/noble
Maria Francesca Raggi	Genoa, 1656–1657	nun
Antero Maria da San Bonaventura	Genoa, 1656–1657	cleric
Maurizio da Tolone	Genoa, 1656–1657	cleric
John Evelyn	London, 1665	noble
Samuel Pepys	London, 1665	bureaucrat
Thomas Allin	London, 1665	naval officer
Simon Patrick	London, 1665	cleric
Henry Oldenburg	London, 1665	scientist
William Blundell	London, 1665	noble
Pierre Ignace Chavatte	Lille, 1667–1669	journeyman sergemaker
Raymundo de Lantery	Cádiz, 1681	merchant
Giovanni Maria Marussig	Gorizia [Italy], 1682	cleric

APPENDIX III

The Plague as Reported in the Barcelona City Council Minutes

The experience of plague was a devastating one. By challenging the certainties of everyday life, it turned familiar worlds upside down and called forth unexpected responses. We have seen how the plague worked a dramatic transformation in Miquel Parets's text. The passages on the travails he and his family suffered during the summer of 1651 represent the only juncture in his lengthy narrative where he abandoned the objective style of the chronicle for a personal engagement with his reader. By a curious coincidence, the same sort of alteration can be found in the Dietari, *or official minutes of the Barcelona city council. Its excursus on the plague of 1651*[1] *is one of the few times in the entire series when the municipal scribe broke its routine impersonality to render a moving portrait of the disintegration of local society under the impact of the epidemic.*

Monday, June 5, 1651. On this day it would be fitting to mention the pitiful travails this wretched and unfortunate city suffers thanks to the sins of its citizens, through the cruel flail of the plague, which has taken hold so thoroughly and with such force that there would be no end were one to try to record the misfortunes, travails, anguish, and woeful deaths which continually take place. For many days now eight or ten carts have travelled throughout Barcelona with the sole purpose of removing corpses from houses, which are often thrown from the windows to the street and then carried off in the carts by the grave diggers, who go about playing their guitars, tambourines, and other instruments in order to forget such great afflictions,[2] the memory alone of which is enough to want to be done with this wretched life, which seems to be worth nothing. These grave diggers stop their carts at a street corner in the city and cry out for everyone to bring the dead from their houses, sometimes taking two from one house, four from another, and often six from another, and after filling their carts they would take the bodies to be buried in a field near the monastery of Jesus called the "beanfield."[3] Apart from these [carts] some forty or fifty stretchers were used to carry those bodies which didn't fit in the carts, and it often happened that the grave diggers would carry dead babies or other children gravely ill with the plague on their backs. The entire city is now in such a lamentable and wretched state that men cannot even remember themselves[4] nor can they imagine the travails they suffer. They recall only that they are Christians, and not even everyone has kept this in mind, for it is certain that the cause of these misfortunes has been our sins, along with our not having mended our behavior before Our Lord raised His hand to punish us. Priests and confessors were missing in almost all the parishes, some having died and others being absent from the city, and as a result monks administered the sacraments in the churches and especially in certain parishes. The need was so pressing that often the priest left the church with the Holy Sacrament (may it be praised), and returned only after having given last rites to fifty or sixty or more persons, and since it was beyond the strength of any person to do so much, he often had to ride through Barcelona on horseback. . . .

In short, the misfortune this city has suffered and still suffers is so great that no single person, no matter how ingenious, can imagine what is suffered in a time of plague, much less describe it and write down all that has happened. It can be said without exaggeration that in Barcelona parents abandoned their children, husbands their wives, and friends fled from each other, so much so that when anyone fell sick with the plague, he or she could rely only on God as a father, friend, or spouse. This pitiable day will finally end and will be remembered in the future as a most cruel plague. May it please God in His infinite goodness and mercy to deliver us from it, and to placate His most just anger and show us grace and mercy. May we mend our sins and shortcomings, for by so doing we will surely obtain His divine mercy.

APPENDIX IV

Barcelona Place Names Referred to in the Parets Chronicle

English version	*Catalan original*
Angel Gate	Portal de l'Angel
Auction Square	Encants*
Booksellers' Street	Llibreteria
Born marketplace	Born
Broad Street	Carrer Ample
Butchers' Street	Carrer dels Tallers
Carmen Street	Carrer del Carme
Curriers' Square	Pla de la Blanqueria*
Hatters' Street	Sombrarers
Hospital Street	Carrer del Hospital
Jesus Street	Carrer de Jesús*
King's Square	Plaça del Rei
Knifemakers' Street	Dagueria
Llansana Street	Carrer d'en Llansana*
Mirrormakers' Street	Mirallers
Moneychangers' Street	Canvis [now Canvis Vells]
Naval yards	Dressanes
Nazareth Street	Carrer de Nazaret [now C. Valldonzella]
New Gate	Portal Nou
New Square	Plaça Nova
New Street	Carrer Nou [de Sant Francesch]
Passageway of the Prison	Devallada de la Presó [now Carrer Llibreteria]
Saint Anne's Square	Plaça de Santana [now Portal de l'Angel]

*No longer in existence.

English version	*Catalan original*
Saint Anthony's Gate	Portal de Sant Antoni
Saint Daniel's Gate	Portal de Sant Daniel*
Saint James's Square	Plaça de Sant Jaume
Saint John's Way	Riera de Sant Joan*
Saint Paul's Street	Carrer de Sant Pau
Sea Gate	Portal del Mar*
Street of Hell	Carrer de l'Infern
Street of the Holy Spirit	Carrer del Sant Esperit*

APPENDIX V

The Uses of Parets

That local scholars have been attracted to the passages on the plague of 1651 within the main body of the chronicle hardly comes as a surprise. One of its earliest readers—Pere Serra i Postius (1671–1748),[1] shopkeeper cum antiquarian scholar and one of Barcelona's other great popular chroniclers—copied lengthy extracts from the Spanish translation of Parets's narrative.[2] Like other early readers of this version of the text, Serra Postius was apparently unaware of the author's identity or even his social background. His numerous editorial interventions included transforming Parets's sober, highly realistic prose into a much more florid style. For better or worse, however, it was Serra Postius's version of the chronicle which most nineteenth-century historians consulted. It served, for example, as the basis of the famed bibliographer Fèlix Torres Amat's[3] description of the chronicle in his 1836 *Dictionary of Catalan Writers*, in which he attributed the chronicle to an "anonymous Barcelonan." The Romantic writer and politician Víctor Balaguer[4] also published a summary of this extract as an appendix to his widely read *History of Catalonia* (1860–1863), although naturally he too knew little about its author.

Thanks to the research of Celestino Pujol i Camps,[5] who not only edited the complete version of the Spanish translation but was also able to locate the original document in Catalan, future students of plague in early modern Catalonia have been able to consult a much more extensive text. Thus, for example, Jordi Nadal and Emili Giralt,[6] in their pioneering study of early modern Catalan demography, frequently cite the Spanish version of the Parets chronicle, as does M. Carreras Roca[7] in his doctoral dissertation on plague in seventeenth-century Catalonia. To my knowledge, however, no one has consulted the original Catalan chronicle for its much more detailed account of the Barcelona epidemic of 1651.[8]

Notes

Full references are given at the first citation of all documents and manuscripts. Printed works are referred to by the author's last name and short title; more complete information can be found in the list of bibliographic references at the end of the book.

Abbreviations

A.C.A.	Archive of the Crown of Aragon, Barcelona
A.H.N	Archivo Histórico Nacional (National Archive), Madrid
A.H.M.B.	Arxiu Històric Municipal (Municipal Archive), Barcelona
A.H.P.B.	Arxiu Històric de Protocols (Notarial Archive), Barcelona
B.C.	Biblioteca de Catalunya (Catalan Library), Barcelona
B.N.	Biblioteca Nacional (National Library), Madrid
B.U.B.	Biblioteca Universitària (University Library), Barcelona
I, II	[B.U.B., Mss. 224–225] Miquel Parets, "De molts successos que han succeït dins Barcelona y en molts altres llochs de Catalunya, dignes de memòria" [original text of the Parets chronicle, upon which this edition is based]
Datos Históricos	J. Ferran, F. Viñas i Cusí, and R. de Grau, "Datos históricos sobre las epidemias de peste ocurridas en Barcelona," in their *La Peste Bubónica. Memoria sobre la epidemia ocurrida en Porto en 1899* (Barcelona, 1907; reprint 1965), pp. 369–625 [official history of bubonic plague in Barcelona commissioned by the city government in 1899]
Deliberacions	A.H.M.B./Consell de Cent II, Registre de deliberacions del consell barceloní, 223 vols. [official minutes of city council meetings]
Dietari	*Manual de Novells Ardits: vulgarment apel.lat Dietari del antich Consell barceloní* (Barcelona, 1892–1975), 28 vols. [published edition of the official daybook of the Barcelona city council]

M.H.E. Miguel Parets, "De los muchos sucesos dignos de memoria que han ocurrido en Barcelona y otros lugares de Cataluña," ed. C. Pujol i Camps, *Memorial Histórico Español* (Madrid, 1888–1893), vols. 20–25 [anonymous Spanish adaptation of Parets's chronicle]

Ordinacions A.H.M.B./Consell de Cent IV, Registre de crides del consell barceloní, 44 vols. [official register of city council decrees]

Introduction

1. Defoe, *Journal*, 81.

2. It is now known that Defoe's uncle, Henry Foe, was indeed a saddler from Whitechapel who survived the London plague of 1665. Bastian ("Defoe's *Journal*") makes the intriguing suggestion that earlier searches for the written sources of Defoe's vision of the epidemic (as exemplified by Nicholson, *Historical Sources*) overlooked his access to significant oral testimonies of the plague, which may explain his having signed the book with the initials "H.F."

3. Gavaldá, *Memoria*, Al. For other studies of this epidemic in Valencia and the Spanish Levant, see Casey, "Crisi"; García Ballester and Mayer Benítez, "Aproximación"; Peset, "Médicos"; Peset and La Parra, "Demografía"; Sanz Sampelayo, "Epidemia"; and Torres Sánchez, "Expansión."

4. A. Domínguez Ortiz, cited in Nadal, *Población*, 38. Pages 38–43 of this study provide a useful account of this contagion, as do Pérez Moreda, *Crisis*, 245–308, and Reher, "Ciutats."

5. The more important studies of the Barcelona plague of 1651 include Camps and Camps, *Pesta*, 71–101; *Datos Históricos*, 494–550; and Viñas i Cusí, *Glànola*. More general references to plague in Barcelona can be found in Biraben, *Hommes*, I, 198–218, and Bruniquer, *Rúbriques*, IV, 319–39.

6. The case of Barcelona was not without parallels elsewhere. Richard Trexler has noted the compatibility in late medieval Florence between acts of communal solidarity (ranging from processions to the delivery of food to pesthouses) and hygienic measures to avoid contagion (*Public Life*, 363), although in another work he characterizes the twin desires to propitiate the divine through processions while avoiding infection as "conflicting" ("Measures," 456–57).

7. The richness of the Barcelona documentation of the plague—remarked, for example, by Jean-Noël Biraben—rests mainly on the fame of its *serca*, or lists of plague deaths noted daily by parish or district. This practice contrasts with the weekly basis of the famous London plague bills, which served as indispensable sources for John Graunt's pioneering seventeenth-century

studies in historical demography, and for writers and diarists like Samuel Pepys and Defoe (Wilson, *Plague in London*, 189–208). Unfortunately, no such bills survive for the 1651 epidemic, as the last comprehensive plague lists date from the 1588–1589 contagion (Biraben, *Hommes*, I, 198–218, and Smith, "Barcelona Bills of Mortality"). This is a serious hindrance to demographic and epidemiological studies, especially given the notorious unreliability of ecclesiastical documentation for the registry of deaths through catastrophic mortality. The principal sources for the 1651 epidemic thus derive from local government records, to which Bruniquer's *Rúbriques* and the *Datos Históricos* provide useful guides.

8. Swanson, "Illusion," 37.

9. For an introduction to the iconography of plague, see Crawfurd, *Plague and Pestilence*, and Mollaret and Brossollet, "Peste."

10. For notations in parish registers, see the 1650 entries from the nearby town of Olot transcribed in Canal, *Vila*, 114; for use of parochial *libri di morti*, see Carlo M. Cipolla's well-known study of the incidents in the Tuscan village of Monte Lupo in 1630 (*Faith*, 99–103). Clerical correspondence from the Catalan city of Terrassa during the plague of 1651 is reproduced in Cardús, *Terrassa*, 211–16; see also Martin, *Jesuit Mind*, 165–70, for letters of Jesuits fleeing from the plague in sixteenth-century France. For a detailed account written by a priest during the plague, see the fascinating diary from Friuli (1598) reproduced in Brozzi, *Peste, Fede e Sanità*.

11. Examples include: the Catalan accounts from the 1590s cited in Nadal and Giralt, *Population catalane*, 38–39n.; the *corregidores*' reports included in Bennassar, *Recherches*, 101–88; the Privy Council letter of 1603 reproduced in Wilson, *Plague in London*, 94–95; and the Milanese reports of 1630 included in *Guerra e Peste*.

12. I know of no comprehensive list of plague tractates, although the bibliographic appendix to Ziegler, *Black Death*, contains a good account for the fourteenth century. The best listing for the early modern period is the lengthy bibliography in the second volume of Biraben, *Hommes*, which should be supplemented by the works of Dorothy Waley Singer ("Some Plague Tractates") and Arnold C. Klebs (*Remèdes contre la peste*). For a brief discussion of early modern Spanish plague treatises, see Carreras Panchón, "Epidemias" and his *Médico y Peste;* Granjel, *Medicina española renacentista* and his *Medicina española del S. XVII;* and Ballesteros, *Peste en Córdoba*. Catalonia boasts an especially rich tradition of plague treatises, including the honor (somewhat disputed) of having the first commentary on the Black Death, written by a physician from Lleida (Veny i Clar, *Regiment*). The neglected history of Catalan plague treatises receives partial treatment in the essay by Joan Veny i Clar cited above, and the introduction by Jordi Rubió to his edition of the converted Jewish physician Lluís de Alcanyiz' *Regiment*, originally published in Valencia around 1490.

Caution is needed when using such documents. For example, not all medical treatises were securely based on actual clinical observation of contemporary epidemics, even when they claimed to be. Many such works reproduced (usually without attribution) passages from earlier tractates. Thus Thomas Lodge's *A Treatise of the Plague*, long favored by plague historians as an eyewitness account of the London epidemic of 1603, has recently been discovered to be a plagiarism of medical works previously published in France (Cuvelier, "*A Treatise*"; and Roberts, "Note"). See Mullett, *Bubonic Plague*, and Slack, *Impact*, 23–24, for sensible remarks on the use of medical treatises as sources for the history of plague.

13. See *Información*. This work is also excerpted in López Piñero and Terrada, "Obra de Porcell."

14. For examples, see Carreras Panchón, "Epidemias," 13, which mentions an 1582 Valencian chapbook about the "great plague" in Cairo, and García de Enterría, *Sociedad*, 213–17. Aragonés, *Lamentación* is one of the few broadsheets to deal specifically with plague in Barcelona. See Griffin, *Crombergers*, 158, for plague sheets published in sixteenth-century Seville.

15. For Protestant references to monks as "public plagues," see Bertelli, *Rebeldes*, 47 and 48. Protestantism was characterized in turn as a "spiritual pestilence" in the 1620s daybook of the Barcelona lawyer Jeroni Pujades (*Dietari*, III, 143). See Martin, *Jesuit Mind*, 95–97, for the French Jesuits' depictions of their Huguenot enemies as (literally) pest-ridden.

16. Xavier Gil drew my attention to the reference to plague in Gonzalo de Céspedes y Meneses' story "La Constante Cordobesa" (in his *Historias Peregrinas y Ejemplares*, 204–8); see Bennassar, *Recherches*, 190, for mention of this theme in another story by Céspedes, "Varia Fortuna del Soldado Pindaro." The playwright Agustín de Rojas also referred to the Seville epidemic of 1599 in a political and moral treatise (*Buen Repúblico*, 6–26), but to my knowledge did not depict it in any of his dramatic works or stories. Such brief and inconclusive mention is typical of the scanty treatment afforded the theme in Spanish imaginative literature, although there are some exceptions to this rule. Especially interesting is an anonymous poem from Logroño (ca. 1599) which contains sustained criticism of the "greedy rich" and the local ruling class in general (Simón, "Otro Romance," esp. 248–49). For the question of fictional representations of plague in early modern Spain, see David-Peyre, "Peste," which unfortunately I have not been able to consult.

17. For the Lisbon chronicle, see Roiz Soares, *Memorial*, 19–39 (my thanks to Fernando Bouza for this citation). The Galician account by the anonymous "cura de Leiro" is excerpted in López Ferreiro, *Historia*, VIII (I am grateful to Juan Eloy Gelabert for bringing this account to my attention). See Serrano, *Anacardina;* and Vega, *Memorias*, 115–28 and 157–76 for the 1649 reports from Málaga and Seville, respectively; Caldera, *Tribunal*, includes a particularly

revealing memoir by a physician of the latter epidemic. The portion of the manuscript by Jeroni de Real dealing with the Girona plague of 1650 has been published as an appendix to Parets, *Pesta;* portions from the Saragossa account by the surgeon José Estiche (*Tratado*) are reproduced in Appendix I of this work, along with excerpts from the Seville accounts. Lantery, *Comerciante saboyano*, 177–79, contains the brief description of the Cádiz contagion of 1681 by a Savoyard merchant. The reports given by the Jesuits can be found in Iglésies, *Pere Gil*, 300–306, and León, *Grandeza*, 537. For the Valladolid texts, see Carreras Panchón, "Dos Testimonios," and Bennassar, *Recherches*, 24, 52, and 190; the Perpignan accounts are reproduced in Pascual, "Mémoires," 219–20, and Torreilles, "Mémoires," 173.

18. The texts of Miquel Onofre de Montfar i Sorts, honored citizen of Barcelona; Jeroni de Real, a noble from Girona; and Joan Francesc Vila, a notary from Vic, are published as appendices to a modern Catalan edition of Miquel Parets's plague chronicle (Parets, *Pesta*). It should also be noted that the Catalan peasant Joan Guàrdia mentioned this epidemic in his diary (Pladevall and Simon, *Guerra i vida pagesa*, 105–07).

19. For the uses by historians of Parets's text, see Appendix V.

20. See *History of the Peloponnesian Wars*, II, 47–54, and III, 87 (although see Parry, "Language," and Poole and Holladay, "Thucydides," for a discussion of the vexing question as to whether this epidemic was really bubonic plague). Another well-known classical account was Procopius's narrative of the so-called "Justinian plague" of 542 AD (*History of the Wars*, II, sec. 22–23).

21. Elisabeth Carpentier remarked on this trend in the introduction to her *Ville*, 7. A good example of this approach is Shrewsbury, *History*, which, while offering numerous quotations from contemporary observers and based on an extremely wide range of sources, nevertheless focuses relentlessly on the clinical aspects and demographic impact of the plague, with virtually no exploration of attitudes and beliefs in relation to the contagion. The interests demonstrated in the most comprehensive history of the plague, Biraben's *Hommes*, are considerably wider; the second part in particular deals with the broader question of responses to pestilence. However, his work resembles Shrewsbury's in the relatively little systematic attention paid to personal (not to mention fictional) accounts of plague experience.

22. Examples include Tenenti, *Senso;* Vovelle, *Mourir autrefois;* Lebrun, *Hommes et la mort*, III, especially 430–35; Delumeau, *Peur en occident*, chapter 3; and Slack, *Impact*, especially chapter 11.

23. M.H.E., vol. 20, xxi (remark by the editor Celestino Pujol i Camps).

24. Reference to plague in the Breslau goldsmith's diary can be found in Vincentz, *Goldschmiede-Chronik*, 52–53 and 116–18 (see also von Greyerz, "Religion," 234 and 238 for details about the author). The Swiss printer and humanist Thomas Platter made constant, if terse, mention of plague deaths in his *Autobiographie;* see also the fascinating retrospective memoir by the

Málaga printer Juan Serrano de Vargas, *Anacardina*. The accounts of the Milanese epidemic of 1576 were written by the carpenter Giambattista Casale ("Diario," 289–309), and by Giovan Ambrosio de' Cozzi (I am indebted to Alessandro Pastore for providing me with a copy of the latter document, edited with the title "Diario di un popolano milanese"). The Aragonese surgeon José Estiche's narrative of the Saragossa plague of 1652 and the contemporary references by the Catalan peasant Joan Guàrdia and the Sevillian shopkeeper Andrés de la Vega are cited in notes 17 and 18. The text by the Bolognese glassworker Michele Risi is found in Malaguzzi Valeri, "Collezionista bolognese"; the 1667 narrative by the journeyman Pierre Ignace Chavatte is discussed in Lottin, *Chavatte*, 144–55. Excerpts from most of these can be found in Appendix I.

Unfortunately, I have not been able to consult the late sixteenth-century Amiens account of the tax official Jean Patte, mentioned in Deyon, "Mentalités populaires," 449, nor the earlier work by the Reims carpenter Jehan Pussot, cited by A. N. Galpern in his *Religions*, 181.

25. This biographical summary derives from scattered documents within Barcelona's Notarial Archive and will be expanded in my introduction to the critical edition of the Parets chronicle currently being prepared by Maria Rosa Margalef.

26. For a brief description of the Barcelona municipal government and a discussion of the question of popular participation in local politics, see Amelang, *Honored Citizens of Barcelona*, 29–33 and 216–22.

27. In the absence of detailed studies of the civic regime of early modern Barcelona, it is difficult to determine the uniqueness of Parets's involvement in local politics. By the late seventeenth century, the lottery for membership in the Council of One Hundred contained some four hundred places for masters from the lesser guilds. Of these, fourteen were reserved for tanners—certainly a respectable sum for a guild whose membership numbered fewer than fifty masters at the time. Given that one tanner served on the council every year, and that at any moment at least one third of the guild members could be nominated to the eligibility lists, the chances of Parets or any other master serving on the council during his lifetime were quite high.

It should perhaps also be noted that the minutes of the council meetings unfortunately do not include any specific information on voting, which was usually carried out by secret ballot. It would therefore be impossible to reconstruct the individual voting record of Parets or any other member.

28. I–II (B.U.B./Mss. 224–25).

It should be kept in mind that these manuscripts are a fair copy and thus probably not written by Parets himself. Unfortunately, I have found no signature or any other writing in the tanner's own hand with which to compare these documents.

29. M.H.E. There is little evidence to indicate when this translation was

written, nor have I been able to locate the original manuscript upon which Pujol i Camps based his edition. However, a note attached to the manuscript copy of a variant Spanish translation (B.C./Ms. 502) suggests that it was completed by 1709. For the history of the different Spanish versions of the Parets manuscript, see Pujol i Camps's introduction to M.H.E., vol. 20.

30. A.C.A./*Reial Cancelleria*, reg. 5584, 31, viceregal edict dated November 16, 1630. The incident of the Milanese poisoners of 1630 achieved remarkable notoriety throughout Europe, including Catalonia. The hideous executions of Mora and Piazza, the two unfortunates held by popular rumor to have caused the plague of 1630 by spreading invisible poisonous powders, were duly noted in both the Barcelona municipal council minutes (*Dietari*, vol. 10, 496–99) and in Bruniquer's *Rúbriques*, IV, 327 and 339. See also Riera and Jiménez Muñoz, "Avisos en España," and their "Dr. Rossell."

31. Both the tanner's endorsement of these official actions and the lack of clerical opposition to the suspension of rogatory processions render all the more unconvincing the simplistic dichotomy which opposes the "rational" behavior of lay bureaucratic elites to "irrational" popular (and clerical) comportment. This dichotomy is posed by Carlo M. Cipolla in his many books on the plague in seventeenth-century Italy, especially *Faith* (see his *Contro un Nemico* for a convenient edition of most of his studies of early modern epidemics).

32. See *Verdadero Conocimiento*.

33. See I Samuel 5–6 for the famed "plague of the Philistines" and Psalms 38 for a description of plague symptoms. Psalms 106 explicitly depicts plague as a punishment for collective sin. Note also the passage in the journal of Pierre Prion, a scribe in a small southern French town, which compares the local plague of 1720 with "la peste de David" (*Pierre Prion, Scribe*, 153).

34. See Montaigne, *Autobiography*, 271; the reference to Sprat's edition of Thucydides is found in Mullett, *Bubonic Plague*, 250. Caldera's memoir of the Seville epidemic of 1649 contains a fascinating comparison of that plague with Thucydides' Athenian narrative (*Tribunal*, 528–34). Medical literature in general drew heavily upon classical metaphors; a frequently cited simile identified the epidemic with the Hydra (see, for example, the introductory epistle by Dr. Josep Fornés i Llorell to Fornés, *Tractatus de peste*).

35. See, for example, the articles in the issues of *Quaderni Storici* devoted to "Calamità, paure, risposte" (55, 1984) and "Terremoti e storia" (60, 1985). William Christian, Jr.'s *Local Religion* and *Apparitions* discuss collective reactions to disasters in Castile and Catalonia. Two recent works, from the perspectives of social psychology and anthropology respectively, include Erikson, *In the Wake of Flood*, and Oliver-Smith, *Martyred City*.

36. The attribution of epidemic disease to collective sin can be found in virtually all early modern works on plague. However, some interesting variants on this theme developed. For example, the sixteenth-century Pari-

sian judge Pierre de l'Étoile (L'Étoile, *Paris of Henry of Navarre*, 74–77) recorded popular rumors to the effect that the plague of 1580 was caused by the king's penchant for amorous liaisons with nuns, while in 1656 the New England Puritan Michael Wigglesworth gloomily believed that his own individual sin could bring judgment, through the form of disease, upon the larger community (*Diary of Michael Wigglesworth*, x). Closer to home, a local broadsheet published in 1589 ends with the author blaming the plague on the lax behavior of local nuns and Dominican monks (Aragonés, *Lamentación*). Note also Gracián's dissenting sarcasm in the *Criticón*, wherein he denied that plague served as a means of divine justice, because it afflicted only the poor and not the rich (cited in Granjel, "Epidemias," 22). Clearly some early modern citizens did not hesitate to identify "collective sin" with the misbehavior of specific individuals.

37. Carreras Panchón, "Epidemias," 14; von Greyerz, "Religion," 228; Torres Sánchez, "Expansión," 124; Ago and Parmeggiani, "Peste del 1656–1657," 194; Pastore, "Tra Giustizia e Politica," 23; Vega, *Memorias*, 116 and 120; Burgos, *Tratado*, 2v.–3v. (also cited in Ballesteros, *Peste en Córdoba*, 126); and Wilson, *Plague in London*, 131.

38. For studies of the "plague origins" of anti vagrancy legislation in late medieval and early modern times, see Pullan, *Rich and Poor in Renaissance Venice*, 219; and Carmichael, *Plague and the Poor*. For Barcelonan examples of the expulsion of vagrants during epidemics, see *Datos Históricos*, 339 and especially 481.

39. See Sanabre, *Acción de Francia*, for background on political developments in mid-seventeenth-century Catalonia.

40. The absence of collective persecution in Barcelona in 1651 contrasts with earlier periods in the city's history, when Jews and other minorities were singled out for repression and often violent attacks by the masses. See López de Meneses, "Consecuencia," and Guerchberg, "Controverse," for a history of anti-Semitic persecutions during the Black Death in Catalonia and Europe, respectively.

The decrease of such violence by the seventeenth century applies only to western Europe. Outbreaks of the plague in eastern Europe continued to give rise to pogroms; hence the fear, during the 1660s, of the Jewish businesswoman and diarist Glückel of Hameln that the discovery of a plague victim would bring fierce retaliation upon the Hanover ghetto (*Memoirs*, 49–56).

41. There is an immense literature on the Italian epidemics of 1630, which witnessed the most dramatic charges of mass poisoning (see note 30). The most comprehensive accounts of the Milanese contagion include the original Ripamonti chronicle, translated into Italian as *Peste di Milano;* Alessandro Manzoni's appendix to chapters 31–37 of his *I Promessi Sposi*, published separately as the *Storia della Colonna Infame;* Nicolini, *Peste e Untori*, and his "Peste del 1629–1632"; and *Guerra e Peste*. For the Florentine epidemic of the

1630s, see Calvi, *Storie;* Lombardi, "1629–1631"; "Curiosi ricordi"; and Sardi Bucci, "Peste del 1630."

Accusations against *untori* were hardly without precedent; indeed, in one form or another they dated back to the Black Death. The sixteenth century witnessed the persecution of *engraisseurs*, or poisoners, in Geneva and Savoy (Monter, *Witchcraft*, 44–49, 115–27) and in Venice in the 1570s and 1630s (Preto, *Peste e Società*, chapter 2; and *Venezia e la Peste, 1348–1797*, 145–46, respectively). There was even a similar incident in Barcelona in 1589 when, according to the local chronicler Narcís Feliu de la Peña, plague was blamed on poisonous powders placed in founts of holy water (*Anales de Cataluña*, III, 216). For contemporary accounts of *untori* in Naples in 1656, see the "Relazione della pestilenza"; Rubino, "Anno 1656"; and B.N./Ms. 1440 [*Varia italiana]*, 356r.–357r.

42. See Calvi, *Storie*, 9–10; as well as the criticism of this work by Edoardo Grendi, "Storia sociale," along with Calvi's reply, "A proposito di *Storie.*"

43. I have consulted the Italian translation by Francesco Cusani (Ripamonti, *Peste di Milano*) of this classic plague treatise, the principal source for Manzoni's depiction of the plague of 1630 in *I Promessi Sposi*.

44. Calvi, "L'Oro," 421, notes the close link between the plague areas and the quarters identified with the Masaniello revolt of the previous decade, the most important popular disturbance in early modern Neapolitan history. Spinola's report is cited on pages 422–23 of this work. A direct connection was also traced between the 1649 plague in Seville and a revolt there in 1652, the "motín de la Feria," in the report within B.N./Ms. 6014 cited by Domínguez Ortiz in his *Sociedad y Mentalidad*, 22–27.

45. The classic study of this question is René Baehrel's "Haine de classe." More recently, other scholars have argued that plague particularly accentuated class consciousness among the elite due to their heightened fear of the poor, who were seen as carriers of the contagion (Slack, *Impact*, 305; and Carmichael, *Plague and the Poor*). It should, however, be noted that some outbreaks of disease gave rise to a temporary suspension of class distinctions. The 1598 diary by the priest Jacopo Strazzolini from Friuli, for example, stresses the lazaretto's function as a place where all were taken and inspected naked without regard to sex, age, and class (Brozzi, *Peste*, *Fede e Sanità*, 33, 40, 42). Once again, what deserves emphasis is the breadth of responses to epidemics and the lack of a uniform pattern of reaction and interpretation.

46. Significantly, while early portions of the chronicle (and other contemporary documents) make ample reference to local rogatory processions and other forms of collective expiation, the fear of contagion soon fostered the privatization of prayer and supplication. This contrasts starkly with religious response within the Moslem societies studied by Michael Dols, where the incidence of plague led to an increase in public religious rites and collective assemblies for communal prayer ("Comparative Communal Responses,"

280). For studies of the use of petitionary prayer and private fasting as Protestant remedies for the plague, see Thomas, *Religion and the Decline of Magic*, 114–24. For the observation that plague fostered individualism instead of community within Catholic societies, see Trexler, *Public Life*, 364; and Calvi, "'Dall'altrui communicatione,'" 190.

47. I am indebted to Giulia Calvi and Geraldine Nichols for bringing this point to my attention. See also Willen, "Women in the Public Sphere," 572, for remarks on the health services women rendered during plagues.

48. The standard reference is to the foreword of the *Decameron*, although see also Dioneo's speech at the end of the sixth day.

49. There are limits to the ethical net which Parets cast. One omission in the application of his moralism concerns his own behavior. The ethics of flight during epidemics, especially by persons holding civic office like Parets himself, was a frequent topic of discussion in plague treatises. However, the tanner brought up the issue only once, in the introductory paragraph to the lengthy section on the sufferings caused by the plague, wherein he briefly argued that no individual should be obliged to witness the anguish caused by the death and abandonment of plague victims.

50. For an extreme statement of this position, see Camporesi, "Cultura popolare."

51. Unfortunately, there are no studies of literacy in early modern Barcelona. Partial data culled from, among other sources, depositions before criminal courts and the Inquisition suggest that, as in much of Europe, the ability to read and write was quite pronounced among urban adult male artisans.

52. Dale Kent emphasizes the practical risks of committing such criticism to paper even in private diaries in her *Rise of the Medici*, 2–3.

53. See Arditi, *Diario*. My reading of this text differs from that of Eric Cochrane (*Historians*, 209).

54. Extracts from the writings of this Burgos patrician (the original manuscript is in the British Library) are reproduced in Brumont, "Coup de grace." The Barcelona city council minutes also registered the parish clergy's flight during the plague; see Appendix III for the complete text.

55. Cozzi, "Diario," and Casale, "Diario." It should perhaps be noted that while both narratives contain fulsome praise of local clerical authorities (especially the singular figure of Archbishop Carlo Borromeo), neither expresses regard for the secular rulers of Milan or their provisions regarding the plague.

56. See Serrano, *Anacardina*, 5v., 6v.–7r., and 15r.–17v.

57. This institution (also known as the Royal Council) was founded in Barcelona in the late fifteenth century and served both as an appellate court and as an advisory board to the viceroy on matters of constitutional prerogative and public policy.

58. Keith Thomas's *Religion and the Decline of Magic* provides the most comprehensive study of the gradual abandonment of traditional magic and ritual practices by a ruling class in early modern Europe.

59. See Deyon, "Mentalités populaires," 455.

60. See *Impact of Plague*, 34 and 241–42. Simon Schama cites the disposition of both learned elites and unlearned commoners to link the visitation of plague with diverse portents as evidence of a "common culture" among "the seventeenth-century Dutch, even if skepticism made headway among the patrician elite" (*Embarrassment of Riches*, 147). That elite and popular cultures agreed more than they differed on basic questions of cosmology and physiology is one of the themes of Piero Camporesi's *Bread of Dreams*.

61. Pladevall and Simon, *Guerra i Vida*, 105.

62. For an interesting recent ethnographic study of beliefs regarding lunar influence on disease in rural Spain, see Guío, "Influjo."

63. See Caldera, *Tribunal*, 508–10. See also J. M. López Piñero's entry on Caldera in López Piñero et al., *Diccionario*, I, 154–55, which characterizes him as a "moderate Galenist" receptive to contemporary medical innovations, including the widely challenged theories of contagion of Girolamo Fracastoro.

64. See Burgos, *Tratado*, 31r.–v., for the influence of the "celestial orb and its stars." Camps and Camps (*Pesta*, 40) list other examples of contemporary belief in astral influence upon plague.

65. Oliva de Sabuco's "Colloquy on Knowledge of Oneself" in her *Nueva Filosofia* (originally published in 1587) notes the influence of the "movement of the heavens" on plague (pp. 138–41).

66. Archivum Historicum Societatis Iesu [Jesuit Archive], Rome/Arag. 21^{II}, 451r.–453v. (report from Huesca dated January 20, 1652).

67. See Pérouas, "L'Univers mental," 36–38.

68. Unfortunately, I have found no information concerning Parets's overall wealth or income, although the fresh hides in his house, noted in his post mortem inventory, suggest that he was actively working as a tanner until his death (A.H.P.B. / Josep Ferrer, *Llibre de inventaris y encans, 1649–1682*, July 7, 1661).

69. I fully agree with Roger Chartier's argument concerning the futility of the "search for a specific and exclusively popular culture" (see "Culture as Appropriation," 235). His general position—that cultural forms mix "corpora" of diverse origins and that "what distinguishes cultural worlds is different kinds of use and different strategies of appropriation" (ibid., 232, 235)—provides a salutary corrective to a priori attempts to infer the characteristics of popular culture in isolation from its specific historical context. For further reflections on this tendency, see the preface to Carlo Ginzburg's *Cheese*, xiii–xxvi; Michel Vovelle's "Intermédiaires"; and Stuart Clark's "French Historians."

70. M.H.E., vol. 24, 401.

71. Carpentier notes the "contrast between the significance [*ampleur*] of the phenomenon [of plague] and the poverty of eyewitness accounts" (*Ville*, 225). Similar remarks can be found in Deyon, "Mentalités populaires," 449. That this could be attributed at least in part to "survivor shock" is suggested by the well-known remark of the Sienese diarist Agnolo di Tura, who wrote in 1348 that many aspects of the plague were so horrible that he could not bring himself to tell of it (Bowsky, *Black Death*, 14).

72. Note, for example, the high pitch of emotion in Saint Ignatius's journal of private prayer, for which he developed a cipher to denote routine outbursts of tears (Loyola, *Inigo*, 6).

73. This rendering can only be found in the Spanish version of the text (M.H.E., vol. 24, 353). Interestingly, Diego de Colmenares used the same theatrical metaphor when referring to the plague of 1599 in his chronicle of Segovia (cited in Martínez, *Segovia*, 128).

Note on the Translation

1. Parets, *Pesta*.

2. See the excerpts in A. L. Rowse, *Sex and Society*.

3. For further discussion of the tendency by members of the early modern Barcelona elite to use Spanish, a socially more prestigious language, see Amelang, *Honored Citizens of Barcelona*, 190–5.

4. See Alcover, *Diccionari*.

The Journal

1. For a general introduction to the city during this period, see Amelang, *Honored Citizens of Barcelona*, 3–23. The following brief account draws heavily upon the two principal studies of the economic and political history of early modern Catalonia, Pierre Vilar's *Catalogne*, I, and John H. Elliott's *Revolt*.

2. This was the fifth largest city in Catalonia and the leading center of population in the Ebro River valley. Lying some ninety miles southwest of Barcelona, Tortosa had well over five thousand inhabitants prior to the plague of 1650.

3. Don Josep d'Ardena i de Sabastida (ca. 1611–1677) was named Viscount of Illa by French king Louis XIII in 1642 for his participation in the defense of Barcelona. A leading political rival of Don Josep de Margarit (discussed later), he eventually served as ambassador to the French court and as a Catalan representative in the negotiations at Münster. For details of the

Tortosa campaign of 1650 and d'Ardena's brief diversionary thrust into Valencia, see Sanabre, *Acción de Francia*, 471–76.

4. These three towns are located immediately south of the border that separated Catalonia and the Kingdom of Valencia.

5. Situated some sixty miles southwest of Barcelona, Tarragona was the sixth largest city in Catalonia and the seat of its archdiocese.

6. Although Parets (like most other early modern plague diarists) provides little information on the symptoms and evolution of the disease in question, it was almost certainly bubonic plague. Biraben, in *Hommes*, gives by far the most detailed medical study of the plague for this period, although it is usefully supplemented by Le Roy Ladurie, "Concept," and the older works by Hirst (*Conquest of Plague*) and Pollitzer (*Plague*).

Parets normally uses the word *mal* (literally, "evil"), a standard term for sickness in early modern Catalan, to denote the plague. Much less frequent are his references to *pesta* or *contagi*. He uses the term *plaga* only once, to evoke plague in the generic sense of collective punishment (along with hunger and war, as with the biblical plagues of Egypt).

7. The high fever produced by plague led many writers to employ the images of fire and immolation to describe its spread. The Jesuit Pere Gil's account of the Barcelona epidemic of 1589–1590 noted that "it slowly caught fire and spread to many people" (Iglésies, *Pere Gil*, 300), while Gavaldá's Valencian narrative referred to plague as a "fire which spreads very quickly" (*Memoria*, sec. xxviii), a usage repeated by the Aragonese surgeon José Estiche during the following decade (*Tratado*, 3r. and 6v.). See Thucydides, *Peloponnesian War*, II, 49; Procopius, *History of the Wars*, II, xxii (15–16); and the prologue to Boccaccio's *Decameron* for other instances of this simile.

8. While Parets's explanation of the primary causes of the plague was providential, he depicts the mechanisms of its transmission—its secondary causes, following Thomistic usage—as natural and "this-worldly." Many of his contemporaries shared this belief, which helps explain why they did not regard secular and religious measures to halt the plague's spread as fundamentally incompatible. In a related vein, Peter Burke has suggested that the growing opposition to plague processions in Catholic Europe during the seventeenth century might have reflected increasing skepticism among theologians regarding the material efficacy of such rituals (*Historical Anthropology*, 231).

9. Placing an embargo on trade was a common preventive measure taken against plague. Most authorities were reluctant to interrupt commerce in this fashion, because it would cause economic damage, as the same measure might be taken against their city were plague found there.

10. In 1640 the city of Barcelona began to mint copper coins to compensate for the shortage of silver currency. Although Parets reckons in *sisens*, or silver

sixpence, note that the city government had to pay in gold ducats. Elliott, *Revolt*, 553–55, contains a table of equivalences for seventeenth-century Catalan coinage.

11. These passes were certificates of good health issued by municipal or viceregal authorities. See Burke, *Historical Anthropology*, 126, for some observations on plague passes as "precocious" modern administrative documents.

12. A village five miles south of Tarragona, it was also known as Vilaseca de Solcina.

13. Both the *Dietari* and the summary in *Datos Históricos* contain detailed chronologies of local votive practices, which in this case followed the customary, well-defined "round" of propitiatory acts. Christian (*Local Religion*) analyzes responses to plague within the more general context of popular religion in early modern Spain; for rogatory processions and formal vows in Barcelona, see page 46 in particular. Other general studies of civic ritual during periods of crisis include: Muir, *Civic Ritual;* Trexler, *Public Life;* Buratti, *Città rituale;* and *Venezia e la Peste.*

14. This is a coastal village about ninety miles north of Barcelona.

15. The Empordà was a *comarca*, or county, in northeastern Catalonia, bordering on France.

16. This was the third largest city in Catalonia, slightly over sixty miles to the northeast of Barcelona. The Girona plague of 1650 is exceptionally well documented, thanks largely to the existence of a detailed diary written by a local patrician named Jeroni de Real. The portion of his daybook which deals with the plague of 1650 has been edited as an appendix to the Catalan edition of the Parets chronicle (Parets, *Pesta*, Appendix III). See also: Busquets, "Població"; Camps and Camps, *Pesta*, 198–204; Clara, "Pesta del 1650"; and Nadal, "Última pandèmia."

17. The Jeroni de Real account, mentioned in Note 16, confirms Parets's suspicion of Dr. Dimas Vileta's acceptance of bribes in return for not declaring the Girona epidemic plague. Predictably, Vileta was one of the many Barcelona physicians removed from eligibility for public office on May 8, 1651, for having abandoned the city. Popular hostility to physicians was intense during the early modern period; see, for example, David-Peyre, *Personnage du médecin*, 45; Ginzburg, "Dovecote"; and Hill, "Medical Profession." Note also that rumors spread that doctors poisoned their patients during plagues, as reported in Calvi, *Storie*, 26–28.

18. The municipal plague magistracy, first created in 1408 (*Datos Históricos*, 374), was a *dotzena* (committee of twelve) appointed from the four estates (elite, merchants, and upper and lower guild masters) represented in the city government. Plague boards throughout Europe tended to draw their membership from a more popular base than was the norm for most municipal institutions. See, for example, Bennassar, "L'organisation municipale," and his *Recherches*, 25 and 59; Nussdorfer, "Civic Salvation"; and Vincent,

Andalucía, 74, which notes the composition of plague boards according to geographic origin (by parish in Loja, and six *barrios* or districts in Seville) during the Andalusian plague of 1596–1602. Similarly, in Madrid in 1648 the plague watch was drawn from groups of citizens elected by their parishes (Granjel, "Epidemias," 20). In fact, René Baehrel ("Epidémies et terreur") has argued that in some circumstances, plague boards served to institute a certain degree of popular control over local elites, and as such provided a memorable precedent for later measures of revolutionary authority, such as what happened in France during the Terror of 1793-1794.

It is only fair to note that the opposite tendency prevailed in many other areas of Europe, where plague magistracies became permanent organs of local government and whose effect was to institutionalize a stricter vigilance over the poor and vagrants. Pullan, *Rich and Poor*, 219, and G. Lotter et al. in *Venezie e la Peste* argue that this is what occurred in Venice; Carmichael, *Plague and the Poor*, notes the same for Renaissance Florence (although see the criticism by John Henderson in the [London] *Times Literary Supplement*, Feb. 20, 1987, p. 189). In any event, such was not the case in Barcelona, where throughout the early modern period the plague *dotzenes* continued to be appointed on a temporary basis and were disbanded when the threat of contagion receded.

19. The word "decided" is repeated in the original manuscript—one of several reminders that Parets's account is retrospective and most likely composed with reference to (now lost) notes kept on a daily entry basis.

20. Both the physician Joan Argila and the surgeon Texidor were sanctioned on May 8, 1651, for having abandoned the city during the plague.

21. In the original, *que nosi burlasen*.

22. The physical isolation of persons who had visited plague-stricken areas or had otherwise entered into contact with the disease was a common prophylactic measure during the early modern era.

23. This organ, also known as the *Audiencia*, was founded in Barcelona in the late fifteenth century. It served both as an appellate court and as an advisory board to the viceroy on matters of constitutional prerogative and public policy.

24. Note that later in the paragraph Parets confuses the name of the judge Camps i Rubí with that of Joan Baptista Gorí, a magistrate who was lynched during a riot in December 1640 (Elliott, *Revolt*, 521).

25. Using stakes to create roadblocks and to mark areas of quarantine was a common practice during epidemics. Unofficial accounts such as Parets's reveal the laxity with which quarantine was often enforced. The violation or actual destruction of the boundary markers—the central motif of Carlo Cipolla's *Faith*—was far from extraordinary.

26. These were the two most popular "disinfectants" of plague. Despite their common usage, contemporary opinion was far from unanimous as to

their efficacy. For example, while London plague orders (Barrett, *Present Remedies*, "Advise" 1r.) recommended the application of vinegar, as did many medical tractates (for example, Rossell's *Verdadero Conocimiento*, 56v.–59r.), the 1598 diary of an Italian cleric criticized it as useless (Brozzi, *Peste, Fede e Sanità*, 46).

27. The reluctance to declare epidemics as the plague was quite typical, as an official declaration of contagion wreaked havoc with commerce and caused serious problems in supplying food to infected cities. Ballesteros (*Peste en Córdoba*, 53), moreover, notes that at times a delay in issuing plague orders allowed the privileged members of local society more time to flee the city. Certain municipal governments, however, earned a reputation for rapid, even ruthless, intervention at the early stages of the outbreak of plague; such was the case of Venice, according to contemporary reports (Brozzi, *Peste, Fede e Sanità*, 11, 15–16, 21, 29).

28. Olot is a small town thirty-five miles to the northwest of Girona. For descriptions of plague there, see Canal, *Vila*, and Camps and Camps, *Pesta*, 161–64.

29. These three flails were commonly mentioned together. See, for example, the following passage from the *Dietari*, dated September 19, 1651: "Now by the grace of God the city is suffering the same three most rigorous punishments that His Divine Majesty had inflicted upon the people of Pharaoh, which are Plague, War, and Famine, the bitterest enemies of humankind" (vol. 15, 312).

30. Food shortages are a constant theme in Parets's chronicle. For example, his account of the 1631 famine (Ms. 224, 35r.–v.) contained harsh criticism of the indifference of the viceroy, the Duke of Cardona, to the sufferings of the poor.

31. This is a measure of grain, equal to 1.7 bushels (Elliott, *Revolt*, 577).

32. At the worst moments of subsistence crises, city governments took over direct control of the baking and distribution of bread. In Barcelona, the *duana*, or building where the grain stocks were stored, was near the present-day Plaça del Palau.

33. Here Parets follows local custom in referring to Catalonia as his *terra* ("land"). Elsewhere he uses the term *pàtria* ("fatherland" or "homeland") in this context, as will be remarked later on.

34. The first direct mention of plague in the official documents of the Barcelona government dates from January 10 (*Dietari*, vol. 15, 91). The Plague Board's January 12 report to the city council on the initial recognition of the disease (*Deliberacions*, vol. 160, 67r.–69v., and published in the *Dietari*, vol. 15, 431–33) contains slight differences in detail from the Parets account, including the fact that the first "plague-stricken" were found "in the house of one Bonora, a bowlmaker and presently the city's Overseer of Weights and Measures." In his diary, the honored citizen Miquel Onofre Montfar i Sorts

referred to December 1650 as the starting date of the Barcelona plague (Parets, *Pesta*, Appendix II), a rumor the tanner notes in the following paragraph.

35. In the original manuscript, "Carrer de Codols" is crossed out, and "Carrer Nou" has been superimposed. This street name refers to the Carrer Nou de Sant Francesch, between the Dormitori de Sant Francesch and the Carrer Escudellers, and not the street now known as "Carrer Nou," which connects the site of the former New Gate with Old Saint Augustine's Square (Balaguer, *Calles de Barcelona*, II, 116–17; and *Datos Históricos*, 497).

36. Langa, while presumably poor, was not a wholly marginal figure in local society, as indicated by his having published in 1641 an eight-page rhymed pamphlet in support of the Catalan cause entitled "A most true relation of the cruelties and impositions of the Count-Duke [the royal favorite Olivares] throughout the Spanish Monarchy, and particularly of the depravity with which he has tried to destroy and annihilate the Principality of Catalonia and the city of Barcelona" (Langa, *Relación*).

37. These were the leading magistrates of Barcelona, selected by lottery from candidates from the city's four official estates. The six posts were divided among three members of the elite (nobles, gentry, honored citizens, [see Note 77], and doctors in law and medicine); one merchant; one *artista*, or upper guild master; and one *menestral*, or lower guild master.

38. This former Dominican convent (the "old angels") served as the first pesthouse. It was located outside Saint Daniel's Gate, within what is now the Ciutadella Park (Ainaud et al., *Catálogo Monumental*, 183–84). It should not be confused with the intramural "convent of the angels" to which the nuns transferred in 1562 when the city bought the old building for use as a pesthouse (Bruniquer, *Rúbriques*, IV, 323). From the beginning of its use as a pesthouse, it was organized following the Italian model; those within it were separated into two groups, the sick and those merely exposed to contagion who were nevertheless obliged to pass quarantine.

39. While there is some uncertainty as to the precise location of the tower of Saint Severus, most contemporary maps depict it as a semicircular bastion at the western end of the Butchers' Street (Galera, *Atlas*, 102; see also the nineteenth-century painting reproduced in Duran, *Barcelona*, I, opposite page 208). The tower of Saint Paul, on the southwestern wall at the end of Saint Paul's Street, looked toward the mountain of Montjuich.

40. Most contemporary plague treatises provided detailed instructions for the disinfecting of houses contaminated by plague. See, for example, Rossell (*Verdadero Conocimiento*, 116r.–120v.) for a list of substances used to fumigate infected houses and clothing.

41. This was a common anti-plague measure, as clothing was believed to carry contagion.

42. Here and elsewhere, Parets's wording reveals that he shared the

widely-held belief that while the ultimate cause of epidemics was divine punishment for collective sin, the actual diffusion of plague was the product of the actions of specific persons.

43. The identification of plague and other forms of illness with dirt and, more generally, pollution, was common.

44. The leading marketplace of early modern Barcelona, it was located immediately to the east of the church of Saint Mary's by the Sea.

45. The royal Hospital of the Holy Cross and Saint Paul, between Carmen and Hospital streets, was the city's largest medical facility.

46. It was widely believed in this era that those who caught and survived the plague were guaranteed future immunity from the disease. Many such survivors thus became plague attendants and nurses, as did Renzo in *I Promessi Sposi.*

47. This refers most probably to Lluís Tristany, a prominent lawyer in mid-century Barcelona.

48. This refers to the parish church of Saints Justus and Pastor, southeast of the City Hall behind Saint James's Square.

49. This short passageway (no longer in existence) connected the Carrer de la Llibreteria and the Plaça del Blat (the present-day Plaça del Angel). It was named for the castle located there, which housed Barcelona's criminal court and prison (Duran, *Barcelona*, I, 185, and III, 292, 530).

50. Barcelona's College of Physicians issued a preliminary report on January 30, 1651 (*Deliberacions*, vol. 160, 100r.v., and *Dietari*, vol. 15, 98), attributing local deaths to the poor diet caused by crop failure. This diagnosis was confirmed in a second report presented on February 10 (ibid., 119r.–120v.). Such "ecological" explanations of the plague, based upon an explicit link between disease and the collective starvation produced by famine, were far from infrequent. Other examples from Spain are reported in Nadal, *Población*, 24–25; Bennassar, *Recherches*, 32–33; and Camps and Camps, *Pesta*, 41. Similar physicians' reports were found in London (Wilson, *Plague in London*, 23) and Naples (Calvi, "L'Oro," 412).

For the much discussed relation between subsistence crises and the plague, see Slack, *Impact*, 75.

51. It is ironic that the health food of today was considered the poison of yesteryear. Parets's strictures against vegetables doubtless reflect the strong prejudice of early modern diet in favor of meat (despite its relative expense, it was seen as an indispensable part of daily fare, even during times of crisis) as well as the belief of his contemporaries that vegetables fostered diarrhea and parasitic infections. Medical opinion was predictably divided on this matter. Rossell (*Verdadero Conocimiento*, 59r.–65v.) recommended eating certain vegetables during the plague, but on the whole seemed to have a higher opinion of meats, especially the leaner varieties.

52. As noted in the Introduction, it was commonly believed that plague

worsened at the full moon and that its future evolution could be predicted on the basis of the moon's activity. Epidemics were seen in turn as portents; hence the malicious glee with which Guicciardini noted the plague that inaugurated the reign of the "barbarian" Pope Adrian VI in 1522 (*History of Italy*, 332–34).

53. This was Barcelona's plenary council, established in 1283 and comprising in the seventeenth century 144 members drawn from all four estates.

It should be kept in mind that Miquel Parets sat on the council at this time. Admitted to the eligibility lists on November 25, 1642 (A.H.M.B./C-VIII, *Insaculacions*, vol. I, 1626–1651, 388r.), his name was drawn in the lottery on November 30, 1650, for a term on the council which ended on March 30, 1653 (*Deliberacions*, vol. 160, extracts). Moreover, on June 20, 1651, his name was also drawn for the Trentenari, an intermediate advisory council of thirty-six members (ibid., 302r.).

54. Parets refers here to the guild militia of Barcelona, in which he also served. The bulk of its patrolling–especially the night watch–was in the hands of apprentices, who also provided the pool of recruits for special levies of the sort mentioned here. For details, see Duran, "La defensa de la ciutat" in his *Barcelona*, II, 109–23.

55. Also spelled *reyals*, these were silver coins worth two shillings.

56. This mountain immediately south of Barcelona was topped by a small fortress overlooking the city.

57. Barcelona's main square and city center housed both the City Hall and the palace of the *Diputació*, or Catalan government.

58. The viceroy was the personal representative of the king and constituted the highest royal authority in Catalonia. Following the revolt of 1640, the viceroy was appointed by the French monarch. However, after the Duc de Mercoeur left the Principality in December 1650, no new viceroy was named until October 1651, when the Marquis de la Mothe received this commission. In the meantime, de facto power in Barcelona lay in the hands of the governor, Don Josep de Margarit, and in particular the personal representative of the French cardinal Mazarin, Pierre de Marca (Sanabre, *Acción de Francia*, 481–517).

59. This refers to the regiment raised by the city of Barcelona for service in the war against Philip IV. For details of its engagements in the early 1650s, see Sanabre, *Acción de Francia*.

60. The portico of the parish church of Saint James was located in the present-day square of that name. The city councillors usually assembled there when going to public festivities. During Corpus Christi Day, the entryway was decorated with palm branches and flowers, and the city's *Aguila*, or eagle figure, danced there before the main procession began (Duran, *Barcelona*, I, 130, 282; II, 543, 570; III, 140).

61. Ecclesiastical privilege allowed church officials to provide refuge for outlaws, which occasioned frequent legal battles when civic and royal officials broke into the churches to seize escaped criminals.

62. The authority of the *juy de proms*, the city's criminal jury, was established in the foundation privilege *Recognoverunt Proceres* of 1283. It consisted of the councillors and twenty-four *prohoms* ("principal citizens") selected from all ranks, and was presided over by the governor or *veguer*, the leading royal law officer.

63. This is Saint Mary's by the Sea, where Parets (who lived on the Volta de Sant Hiacinto facing the present-day marketplace of Saint Catherine's) was a parishioner.

64. This area of the city between the central thoroughfare known as the *Rambla* and the southwestern walls centered around Saint Paul's Street. It was one of the more impoverished and sparsely populated districts of the city.

65. The patron saint of Barcelona, whose shrine was located on Montjuich (Ainaud, *Catálogo Monumental*, 225). A "specialist" saint, her intercession was invoked during communal disasters, especially during droughts, when her body was carried down from Montjuich to the city for extensive rogatory processions. For details concerning this devotion, see Duran, *Barcelona*, II, *passim*.

66. The obligations of this important official, charged with enforcing plague regulations both in the pesthouse and throughout the city, were spelled out March 14, 1651, in the *Deliberacions* (vol. 160, 157r.–158r.).

67. This extramural Franciscan monastery (founded in 1427) was located until 1835 near the present-day Carrer d'Aragó, between the Passeig de Gràcia and Pau Claris (Ainaud, *Catálogo Monumental*, 225–26; Barraquer, *Casas de Religiosos*, I, 483–90; Duran, *Barcelona*, II, 71). The history of this monastery is summarized in B.U.B./Ms. 191, P. Serra i Postius, "Historia ecclesiàstica de Catalunya" (vol. 6, 46r.–53v). For the city's request to use the monastery as a pesthouse, see *Deliberacions* (vol. 160, 99r–v., January 28, 1651); for the order to transfer the sick from *Angels Vells* to Jesus, see the *Dietari*, vol. 15, 99–101 (February 2–5); and for the internal organization of the new pesthouse, see the *Deliberacions* (vol. 160, 104r.–107r., published in the *Dietari*, vol. 15, 455–57).

68. This village immediately to the southwest of Barcelona is now incorporated within the city limits.

69. "Different sorts of plague boils," or in the original, *bonys y carbunclos*. "Carbuncle" is not used here in the strict diagnostic sense of the malignant cutaneous pustule caused by *bacillus anthracis* but instead as a general reference to a plague boil. Detailed descriptions of the differences between the two and another type of boil, referred to by Parets as a *vèrtola* on f. 39v., can be found in a treatise by the local physician Josep Mas (*Orde Breu*, 80r.–87v.),

Rossell (*Verdadero Conocimiento*, 107r.–115v.), and Burgos (*Tratado*, 17v.–24r.).

70. Girls (usually dressed in white gowns, symbolic of their purity) were a prominent feature of local rogatory processions. Note that Parets refers to penitents as *pelegrins* ("pilgrims").

71. This is the first reference to the prohibition of collective devotions due to fear of contagion.

72. Part of the Holy Week festivities in Barcelona was the erection and decoration of numerous replicas of the Sepulchre of the Resurrection. These altars could be found throughout the city, especially in the chapels of confraternities.

73. Calabrian saint (ca. 1416–1507), founder of the Minims, canonized in 1519, and renowned for his extreme austerity. A house owned by the church bearing his name located near the New Gate had been used as a pesthouse during the plague of 1589 (Mas, *Orde Breu*, 67v.) and served as a refuge for plague-stricken women beginning on July 6, 1651 (Bruniquer, *Rúbriques*, IV, 332). Details on the elevation of this saint to patron saint of Barcelona can be found in: *Datos Históricos*, 509–11; *Deliberacions*, vol. 160, 168r.–v. (March 22, 1651); *Dietari*, vol. 15, 119–22, 127–28; and *Ordinacions*, vol. 33, 1r.–2r. (which gives the April 14, 1651, decree adding St. Francis de Paul to the city's patron saints).

Earlier plague devotions in Barcelona focused on saints Roch and Sebastian: Duran, "Les devocions dels Consellers," in *Barcelona*, II, 518–23; and Christian, *Apparitions*, 132, 141. The Calabrian saint was probably chosen in March as a patron because of the proximity of his feast day (April 2); his earlier service as a spiritual intermediary during the plague of 1589 (Feliu de la Peña, *Anales de Cataluña*, III, 216); the general recognition of his thaumaturgic powers; and his fame as a model penitent, based on his rigorous fasts (he never ate meat, milk, and eggs) and self-flagellations. Note also his invocation during plague times in other areas of early modern Spain, including Perpignan (Torreilles, "Mémoires," 215), Málaga (Serrano, *Anacardina*, 8v.–9v.), Olot (Canal, *Vila*, 50), Girona (Clara, "Pesta del 1650," 175), Mallorca (Vaquer Bennasar, "Peste de 1652"), and New Castile (Christian, *Local Religion*, 28–29, 42–43, and 97).

74. Civic vows were a familiar part of religious life in Catholic Europe during the early modern era. See Christian (*Local Religion* and *Apparitions*) for detailed accounts of this practice in medieval and early modern Spanish cities. Some early modern civic vows led to the establishment of permanent devotions, like the annual plague procession instituted in Venice; for details see Pullan (*Rich and Poor*, 249–51, 314–25) and Muir (*Civic Ritual*, 214–16).

75. This refers to Saint Raymond of Penyafort, whose feast day is celebrated March 15.

76. The church and monastery of the Minims was located on the Carrer de

Sant Pere *més alt*, next to the present-day Palau de la Música, or Municipal Concert Hall.

77. The ruling class of Barcelona was divided into four ranks. These included, from top to bottom: aristocrats proper and titled nobility; gentry; "honored citizens" (ennobled rentiers occupying a rank slightly below that of the gentry); and doctors in medicine and law, who enjoyed certain non-hereditary privileges of nobility. For a detailed study of the urban elite, see Amelang, *Honored Citizens of Barcelona.*

78. Artisans, or *artistes*, were members of the upper guilds, such as notaries and surgeons. Although Parets does not mention it explicitly, *menestrals*, or members of the lower guilds, such as himself, would also have participated in the procession.

79. This is the building housing the present-day municipal administration in Saint James's Square. For its history, see Duran, "Casa de la Ciutat," in his *Barcelona*, I, 279–323.

80. This convent, founded in 1358, was demolished in 1877. Located near what is now known as the Via Laietana, it housed a small community of nuns of the Augustinian order.

81. I have been unable to identify the present-day whereabouts of this votive painting.

82. The councillors serving during the plague year were Hiacinto Fábregues, honored citizen (who died while in office on April 12, 1651, and was replaced by the lawyer Francesch Vila); Dr. Francesch Matheu, physician; Joan Carreras, gentleman; Josep Rubió, merchant; Joseph Paissa, royal notary; and Miquel Llargues, silversmith.

83. Both the dating and wording of this passage contradict his earlier account of the rain rogations of Saint Madrona. Perhaps Parets wished to indicate merely that it hadn't rained enough in March.

84. This Italian port was one of the leading depots for the bulk shipment of grain in the Mediterranean.

85. Saint Severus was bishop of Barcelona during the seventh century. His relics were "discovered" in Sant Cugat in 1226 and transferred to Barcelona in 1405. The celebration of his feast on April 30, 1651, is described in the *Dietari*, vol. 15, 133–34. For references to local devotion to St. Severus, see Duran, "La retaule de St. Sever, exiliat i repatriat," in his *Barcelona*, III, 338-47, and other references in [illegible] and III, *passim*.

85. The procession of Corpus Christi was the most spectacular and elaborate of street celebrations in seventeenth-century Barcelona, and thus often served as a model for other parades, both religious and secular. See Duran, "La festa del Corpus," in his *Barcelona*, II, 529–71. Reference to this procession in 1651 can be found in the *Dietari*, vol. 15, 159 (June 8).

87. The Marquis of Aguilar (1602–1685) was named governor of Catalonia

in 1641 and ambassador from the *Diputació* (see Note 95) to France at the beginning of the same year. One of the leading politicians of the period, he was closely linked to the faction headed by Pierre de Marca. For details, see Sanabre, *Acción de Francia*.

88. Saint Severus was one of the patrons of the guild of the *teixidors de llana* ("wool weavers").

89. This edict was registered on April 6, 1651 (*Dietari*, vol. 15, 123). See *Datos Históricos*, 512 and 532, for similar instances.

90. Priests tended to wear shorter cassocks during epidemics to avoid the contagion allegedly transmitted by street dust.

91. Saint Mary's by the Sea.

92. This is confirmed by a fascinating diary written by an Italian cleric during the Friulian plague of 1598, which vividly conveys the sense of fatigue caused by constant travel within the city (Brozzi, *Peste, Fede e Sanità*).

93. The mortality rate during plague periods of regular (as opposed to secular) clergy was often quite high. The chronicler Feliu de la Peña (who based this volume in part on Parets's manuscript) included a partial list of monks and friars who died while tending the sick during the Barcelona plague of 1651 (*Anales*, III, 317–18).

94. No choice, that is, but to make an official declaration of plague. This was done in Barcelona in the plague orders of April 21, 1651 (*Ordinacions*, vol. 33, 2r.–8r.), although the orders themselves were not published until early 1652 (*Advertiments Convenients*).

95. This was the name for the three members of the *Diputació* (also known as the *Generalitat*), the standing commission of the three Estates of the Catalan parliament. They were among the most visible figures within Catalan politics, a result of their responsibility for overseeing national finances and ensuring that the viceroy and other members of the royal government did not violate any of Catalonia's constitutional traditions and prerogatives. The *Diputats* were selected by lottery for three-year terms, along with three *Oidors de Comptes* (Overseers of Accounts).

Flight from epidemics appears to have been standard procedure among members of both Catalan and royal administrations; the *Diputats* abandoned Barcelona under the threat of plague at least thirteen times between 1439 and 1651, according to the *Datos Históricos*. In fact, Bruniquer noted that the threat of plague in 1560 could not have been very serious, as the Royal Council did not leave the city (*Rúbriques*, IV, 323). For details concerning the negotiations and correspondence between the city authorities and the Royal Council and *Diputació* in 1651, see *Datos Históricos*, 515, 525–30. The *Audiencia* returned to Barcelona on August 12, 1651 (*Dietari*, vol. 15, 175); for the *Diputats*' refusal to return during the siege of 1652, see II, 52r.

96. Terrassa is a town twenty miles northwest of Barcelona. For the

transfer of the *Diputació* there during the plague of 1651, see Cardús, *Terrassa*, 160–217.

97. This is the seigneurial palace on the left bank of the Besós River within the municipality of Santa Coloma de Gramenet, near the former monastery of Saint Jerome of the Myrtle. Rebuilt in the sixteenth century by the prominent Cardona family, in 1561 it passed into the hands of the Cassadors, well-known members of the Barcelona patriciate. For details, see Duran, "La Torre Pallaresa, una residència senyorial," in his *Barcelona*, I, 716–767.

98. In original, *la cosa se adobaria*, a colloquialism often used in this period. See, for example, its frequent appearance in the lengthy early seventeenth-century chronicle by the Valencian priest Pere Joan Porcar (*Dietari*).

99. The reference is to a bridge at the extramural settlement of Sant Martí de Provençals, now incorporated within Barcelona's city limits.

100. *Escorxador*, or slaughterhouse, in the extramural village of Sants, immediately to the southwest of Barcelona.

101. The *llacuna* (lagoon) was a swamp to the northeast of Barcelona, presently the site of the Barcelona suburb of Poble Nou.

102. The *revenedors* ("retailers") acted as middlemen between farmers and consumers, specializing in the resale of fruits and vegetables in the Born marketplace (Duran, "La fruita," in his *Barcelona*, II, 435–37, and I, 480, 502).

103. Criticism of judges is a constant theme throughout the Parets chronicle. See, for example, his account of the lynching of several judges during the Corpus Christi Day riot of 1640 (Ms. 224, 53r., June 7, 1640).

104. Jean-Gaspard-Fernand Marchin (1601–1673), Count of Marchin, French officer, named lieutenant-general of the French army in Catalonia in 1647. After being ordered arrested by Mazarin in 1650 as an ally of the Prince de Condé, he was released, and he returned to Catalonia in April 1651 (Sanabre, *Acción de Francia*, 447–48, 484).

105. The Prince de Condé was one of Mazarin's principal opponents during the *Fronde*, or civil war in France during the later 1640s and early 1650s. For the influence this struggle had on events in Catalonia, see Sanabre, *Acción de Francia*, chapter 16.

106. A town eighteen miles north of Barcelona.

107. Fourth largest city in Catalonia, approximately 100 miles due west of Barcelona.

108. A town seventy-five miles west of Barcelona.

109. The charity officials, or the *almoyna*, were organized into an agency for charitable relief by the Cathedral chapter of Barcelona. For a recent study, see Fatjó, "Instrument."

110. "upper cells . . . outer cells and the patio . . . dungeon": *tarrat . . . cambra . . . coral . . . Judeca.*

111. The division of the city into four, five, and eventually six quarters for closer supervision was a crucial step toward the increasingly localized

organization of resistance to the plague. For details see *Deliberacions*, vol. 160, 187r.–v. (April 1, 1651), and *Datos Históricos*, especially pages 512, 526, 546–48, 555–56.

112. A valley immediately south of Barcelona, between the mountains of Montjuich and Pedralbes.

113. The motif of empty streets commonly recurs in descriptions of plague-stricken cities. Other examples include: Procopius, *History of the Wars*, xxiii, 17; diverse testimonies cited by Wilson, *Plague in London*, 150–51; and Defoe, *Journal*, 104ff.

114. Here begins Parets's criticisms of the local administration of plague relief.

115. The order of immuration dates from July 29–30, 1651, and can be found in *Ordinacions*, vol. 33, 18v.–20r. Casale, in "Diario," 298–308, is one of the few first-person accounts written by a survivor of strict immuration (in his case, during the Milan plague of 1576); see Appendix I for a translation of part.

116. The Marcús Chapel stood at the intersection of the Carrer de Montcada and the Carrer dels Corders. The incident of the thief is related in the *Dietari*, vol. 15, 160–61 (June 13, 1651).

117. The University was located on the *Rambla* near the present-day Font de Canaletes, and the building that formerly housed the Tridentine Seminary is now the Casa de la Caritat ("Charity House") on the Carrer de Montalegre. The Church of Nazareth was located nearby in the upper part of the district known as the *Raval*, or "suburb" (Ainaud, *Catálogo Monumental*, 38, 228, 235). (See pp. 81–88 for its use as a pesthouse during the later stages of this epidemic.)

118. The latter refers to the thorough cleaning of body and personal effects described on pp. 74–75.

119. Honorific post linked to the royal courier service established in the thirteenth century. Interestingly, Barcelona's guild of couriers was located in the Marcús Chapel (Duran, *Barcelona*, II, 493–501).

120. Plague jurisdiction in Catalonia was almost exclusively municipal in character, as in much of the rest of Europe. For example, the English Parliament passed only one act of plague legislation during the entire early modern period (Wilson, *Plague in London*, 59). While the most solid legal basis for this prerogative dated from a royal privilege of 1510 (published in Capmany, *Memorias Históricas*, doc. 440, pp. 639–40), significant precedents dated from the fourteenth century (*Datos Históricos*, 411n., 430, 484). *Deliberacions* (vol. 166, 48r.–54r.) also thoroughly discusses the legal case for the city's jurisdiction, which was strongly disputed by the viceroy in 1631 (see Bruniquer, *Rúbriques*, IV, 328, 339; and Friedman, "Public Health"). For studies of municipal responses to plague elsewhere, see: Alexander, *Bubonic*

Plague; Andel, "Plague Regulations"; Bennassar, "L'Organisation municipale" and *Recherches;* Charlier, *Peste à Bruxelles;* D'Amelia, "Peste romana"; Fortea, *Córdoba*, 173–219; Gaffarel and Duranty, *Peste à Marseille;* Hildesheimer, "Prévention"; La Parra, *Tiempo de Peste;* Montemayor, "Ciudad"; Nussdorfer, "Civic Salvation"; and especially Paul Slack's excellent "Metropolitan Government" and *Impact*, part III.

121. This scandal is referred to in the *Dietari*, vol. 15, 160 (June 7, 1651). Accusations of sexual misconduct, particularly in pesthouses, figure frequently in plague narratives. For example, the Jesuit Pere Gil was surprised to find so few "crimes of the flesh and robberies" during the plague of 1589–1590 (Iglésies, *Pere Gil*, 303), although a false plague healer was executed for rape (among other things) during that epidemic (*Datos Históricos*, 443). For references to sexual laxity during plague as occurred elsewhere, see: Haton, *Mémoires*, I, 226, for northern France in 1561; Calvi, "L'Oro," 433–34, for Naples in 1656; and Pastore, "Tra Giustizia," 229–31, for Genoa in 1656–1657.

122. The treatise in question is Rossell's widely circulated *Verdadero Conocimiento* (56r.). For biographical information on this thoroughly disagreeable character, see: Amelang, *Honored Citizens of Barcelona*, 41–43; *Datos Históricos*, for his refusal to undertake a plague inspection in 1599 (449n.) and his subsequent willingness to do so in 1629 (469); and Riera and Jiménez Muñóz, "Doctor Rossell."

123. For the history of this often-repeated proverb, see Veny i Clar, *Regiment*, 32, n.; and Biraben, *Hommes*, II, 160–61. Other examples can be found in Burgos, *Tratado*, 62r.; and Granjel, *Medicina española del Renacimiento*, 208 (citing Diego de Torres). Not all early modern writers agreed with this recommendation, however. While Leon Battista Alberti (whose mother died of plague) strongly justified flight, the civic humanist Coluccio Salutati attacked those who abandoned their posts during times of pestilence (see Alberti, *Family in Renaissance Florence*, 126–27; McClure, "Art of Mourning," 444).

124. Interestingly, Parets uses the term *conversasió* ("conversation") to denote personal contact among friends. The plague's impact on everyday sociability was one of the consequences most bitterly remarked upon by contemporaries. For example, the Protestant theologian Richard Baxter noted that during the sickness "every man was a terror to another" (*Autobiography*, 196), while the physician Nathaniel Hodges spoke of the "consternation of those thus separated from all society" (Dyer, "Influence," 320). Similar comments can be found in : Camps and Camps, *Pesta*, 29; Canal, *Vila*, 84–85; Reher, "Ciutats," 105n.; Slack, "Metropolitan Government," 73; and Thomas, *Religion and the Decline of Magic*, 8.

125. Complaints about plague nurses were common during this era. See,

for example, the bitter remarks of the English dramatist Thomas Dekker cited in Wilson, *Plague in London*, 67–68.

126. Parets's obvious affection for children was not limited just to his own family, although this passage was certainly colored by his personal experience during the sickness, as will be seen later in this section. This and other paragraphs provide further evidence against the argument of Philippe Ariès, Lawrence Stone, and others that lower-class family life prior to the modern era lacked warmth and emotional response—unless, of course, it were argued that Parets and other artisans belonged more properly to the middle instead of the popular classes. For criticisms of their position, see Thompson, "Happy Families"; Macfarlane, "Family"; Herlily, *Medieval Households;* and Pollock, *Forgotten Children*.

127. *llet munyda* ("squeezed milk"), breast milk given or sold by lactating women. Interestingly, Parets's account—in many ways a revealing description of early modern nursing practices—makes no reference to the consumption of animal milk, even in such desperate circumstances. Thomas Platter also remarked his early consumption of cow milk as unusual (*Autobiographie*, 21).

128. Early modern wet nurses were frequently criticized as negligent and uncaring of their charges (Pollock, *Forgotten Children*, 50); other contemporary diarists complained of their exorbitant cost (Modena, *Autobiography*, 97 and 133). See also the remark of the shopkeeper Andrés de la Vega concerning the abandonment of children during the Seville epidemic of 1649 (*Memorias*, 124).

129. *escorxada* (literally, "skinned").

130. Mas (*Orde Breu*, 37v.–38v., 92r.–93v.) contains some interesting remarks on the special measures needed to protect pregnant women during the plague.

131. Here Parets expresses the fear that French immigrants might have been Huguenots instead of Catholics. Seventeenth-century Barcelona housed a large community of both resident and transient Frenchmen; it is estimated, in fact, that in the 1630s fully one fifth of the city's population was born north of the Pyrenees or had parents who originated there (Giralt, "Colonia").

132. The tanner begins here the tale of his own journey outside the city, which can be usefully compared with the similar account of the honored citizen Miquel Onofre Montfar i Sorts (see Appendix II in the Catalan edition of Parets, *Pesta*). The artisan's narrative makes clear that flight was not limited just to members of the urban elite.

Note also in this connection the list of city officials ordered on May 8, 1651, to appear or face removal from eligibility for public office (*Deliberacions*, vol. 160, 238r.). Those who appeared can be broken down by estate, as follows: both of the gentlemen and aristocrats cited appeared; none of the four members of the mixed group of honored citizens, lawyers, and physicians cited appeared; four of the six merchants presented themselves; only five of the eleven upper guild masters (mostly notaries) showed; and a mere five of the sixteen lower guild masters cited appeared. The final list of sanctioned officials can be found in the *Dietari*, vol. 15, 189–201 (September 19, 1651).

133. Towns immediately to the south of Barcelona on the Llobregat River.

134. Village on the northeast outskirts of Barcelona, now incorporated within the city limits.

135. Caterina Parets, daughter of Pere Alaver, butcher and citizen of Barcelona, married Miquel Parets the elder in 1606. Widowed in 1631, she died in 1670.

136. Brother of Elisabet Mans, Parets's second wife, whose death is related below (see Note 139). The family farm was most probably located near the Creu d'en Mans, a cross fixed until the nineteenth century at the intersection of the present-day Travessera de Les Corts and the Carretera de Sarrià (Duran, *Barcelona*, I, 618).

137. Interestingly, the *Dietari* of August 12, 1651, noted that "the enemy army advanced along the slopes of the mountains of Pedralbes and Sarrià, until it reached the farm known as that of Mans of Sarrià" (vol. 15, 175). See Feliu de la Peña, *Anales de Cataluña*, III, 318–29, and Sanabre, *Acción de Francia*, 503–54, for details of these military operations.

138. *frenètichs* ("frenzied," "frantic"). References to the delirium and "distorted imagination" produced by the plague fever were common in commentaries on this sickness; examples range from Procopius (*History of the Wars*, xxii, 21) to the printer Juan Serrano de Vargas's references to cases of delirium and madness during the Málaga plague of 1649 (*Anacardina*, 12v.). The Valencian cleric Pere Joan Porcar employed the term to refer to both a suicide and a "bewitched woman" (*Dietari*, 231 and 159, respectively). The latter usage serves to remind us that some early modern writers (including Bodin and the Jesuit Martín del Río) attributed the plague to the machinations of demonic spirits (Thomas, *Religion and the Decline of Magic*, 472).

139. Elisabet Parets i Mans (ca. 1621–1651), second wife of Miquel Parets, whom she married in 1637 (marriage contract in A.H.P.B./Josep Galcem, *Manual de Capítulos Matrimoniales 1627–1650*, no. 67). Daughter of Joan and Eulària Mans, peasants from Sarrià, she died intestate and, as this passage suggests, apparently left no inventory.

Not all plague diarists commented on the deaths of their spouses; certainly very few did to the extent and with the depth of emotion of Parets. For example, the Bolognese bricklayer Gaspare Nadi's one-sentence entry on the death in 1468 of his second wife from plague (*Diario*, 63); the Valencian patrician Jeroni Soria's unemotional annotation of the same in 1558 (*Dietari*, 168); and the passage by the Buckinghamshire clergyman on the death of his wife and children, cited by Shrewsbury (*History*, 424–25) were much more terse and noncommittal.

140. This large Dominican monastery in Barcelona was located immediately in front of Parets's house in the *Ribera* quarter of the city. Elisabet Mans's private devotional proclivities were apparently closely linked to this church; not only did she confess there but the presence at her deathbed of a Rosary candle suggests that she was a member of the devotional confraternity of that name housed there, the largest lay spiritual body in early modern Barcelona.

141. Notarial activity declined abruptly during plague years, largely due to the flight or death of the notaries themselves. As a result, normal requirements in regard to the minimum number of witnesses and the like were relaxed during epidemics, as noted in Anglada, "Testament." However, one should also keep in mind Alessandro Pastore's observation that notaries were one of several occupational groups that often derived substantial advantages during plagues ("Peste e società," 868–69). See also his revealing analysis of notarial practice in Bologna during the plague of 1630 ("Testamenti"), whose conclusions contrast with what Richard Emery found in his study of notaries and the plague in mid-fourteenth-century Perpignan ("Black Death").

142. Anna Maria Magdalena Eulària Parets i Mans (May 3, 1650–May 16, 1651), baptized May 5, 1650, in Saint Mary's by the Sea.

143. Joan Francesch Antich Gabriel Parets i Mans (March 22, 1647–?), baptized March 25, 1647.

144. Fernando Miquel Josep Benet Parets i Mans, "Miquelo" (March 21, 1639–May 9, 1651), baptized March 23, 1639, in Saint Mary's by the Sea.

145. The term in the original is *glànola*, which may refer to the disease known at that time as *garotillo*, or diphtheria. For the difficulties in retrospectively diagnosing plague symptoms, see Poole and Holladay, "Thucydides," and Carmichael, *Plague and the Poor*, 10–26, 85–89.

146. Josep Hiacinto Francesch Jaume Parets i Mans, "Josepo" (August 16, 1640–May 1, 1651), baptized August 19, 1640, in Saint Mary's by the Sea, where he was eventually buried.

147. Parets's lament recalls the poem his near-contemporary Ben Jonson wrote after the death of his son from plague:

> Farewell, thou child of my right hand, and joy,
> My sinne was too much hope of thee, lov'd boy;
> Seven yeeres tho' were lent to me, and I thee pay,
> Exacted by thy fate, on the just day.
> O, could I loose all father, now. For why
> Will man lament the state he should envie?
> To have so soone scap'd worlds, and fleshes rage,
> And, if no other miserie, yet age?
> Rest in soft peace, and, ask'd say here doth lye

BEN. JONSON his best piece of poetrie.
For whose sake, hence-forth, all his vowes be such,
As what he loves may never like too much.

(Parfitt, *Ben Jonson*, 48; for the circumstances surrounding its composition, see Wilson, *Plague in London*, 112–13 and Riggs, *Ben Jonson*, 93-97.)

While the laments of fathers for the death of their sons were regrettably a commonplace of late medieval and early modern literature, few descriptions match that of Parets for emotional intensity and expressive quality. Compare, for example, the 1374 account by the Florentine Giovanni Morelli of the death of one of his sons (cited in Brucker, *Renaissance Florence*, 146–47); it is written in the third person and registers no overt feeling. For other examples of literary records of parental grief during this period, see the brief entry by the Sienese Agnolo di Tura del Grasso on the death of his children during the Black Death (cited in Carpentier, *Orvieto*, 7), and the longer narrative by the Venetian humanist Antonio da Romagno on the death of his children (including his favorite son) during the 1400 plague at Feltre (*Venezia e la Peste*, 53). For other examples of parental grief during the later Middle Ages and early modern era, see: Macfarlane, "Family, Sex and Marriage"; Alberti, *Family in Renaissance Florence*, 126–27 (which refers to the letter from Alberti to his tutor commenting on the death of the former's mother during plague); Banker, "Mourning a Son"; McClure, "Art of Mourning"; King, "Inconsolable Father"; Trexler, "In Search of Father" (later expanded into pp. 159–86 of *Public Life*); and the 1631 diary entry by the Pescian notary Giuliano Ceci (cited in Brown, *Shadow of Florence*, 57).

148. Gabriel, the only child of Parets to survive the plague, professed as a Discalced Augustinian in 1664 and entered the local monastery of Saint Monica. He was still living as late as 1693, although all trace of him afterwards has been lost. The records of his 1664 examination for entrance into the Augustinian order can be found in A.C.A./Monacals, Hisenda, 648, no. 47.

Like many of his fellow survivors, Miquel Parets re-wed shortly after the end of the siege of 1652. On January 3, 1653, he married the (also widowed) Marianna Corbera. While their marriage contract apparently has not survived, their union was noted in the diocesan registry (Cathedral Archive, Barcelona, *Esposalles*, vol. 86, 66v.). The couple later went on to have four children, two of whom survived the tanner's death in 1661.

149. The exact number of plague deaths is not known, due to the lack of reliable records. The high estimate of thirty thousand deaths was also the figure given by Feliu de la Peña and it is indeed likely that he took it from Parets. For further discussion, see *Datos Históricos*, 526; and Nadal and Giralt, *Population catalane*, 44.

150. From his vantage point in Girona, the diarist Jeroni de Real also noted this urgent political reason for the early declaration of the end of the plague

(Appendix III of Parets, *Pesta*, under August 7, 1651). The same connection is mentioned in the *Dietari*, vol. 15, 171 (July 27, 1651).

151. Here as elsewhere in his text, Parets reveals himself to be a strong defender of the Catalan cause during the war against Spain. Interestingly, the diary of the peasant Joan Guàrdia also noted that Barcelona would not have surrendered in 1652 had it not been for "disaffected Catalans" (Pladevall and Simon, *Guerra i Vida*, 73).

152. Parets uses the term *pàtria*, which could refer either to his native city of Barcelona or to Catalonia as a whole.

153. Candidates for public office in Barcelona were chosen by lottery, in a complicated system known as the *insaculació*. The councillors nominated representatives from all four estates for membership in the Council of One Hundred, whose composition was fixed by a series of royal privileges dating from the later fifteenth century. Following scrutiny of the nominees by a committee chosen by the council, the candidates' names were placed in bags and were drawn in a public lottery held each November. The purpose of this procedure was to provide a neutral means of access to civic office and thus prevent any recurrence of the factionalism and bitter internal disputes which had so weakened the municipal administration during the later Middle Ages.

154. Once again, it should be kept in mind that Parets is an eyewitness to this discussion, as he continued to serve on the council until November 1653.

155. The municipal government of Barcelona habitually retained lawyers to provide advice (in the form of written opinions) on a wide variety of legal and constitutional issues. For details, see Amelang, "Honored Citizens and Shameful Poor," 173–74.

156. Parets clearly approves of this measure. Ms. 225, 91v.–92r. describes in detail the November 1651 attempt to reinstate the expelled members of the city council.

157. Both were streets located in the northeastern corner of the city, and both were demolished during the construction of the Ciutadella fort beginning in 1716. Shortly thereafter the local mercer, Pere Serra i Postius, wrote a nostalgic evocation of the gardens and houses of the *fusina* in his fascinating dialogue *Lo Perqué de Barcelona*. The term itself derives from the Latin *officium*, and generally was used to indicate an area given to smelting or other work with metal forges.

158. *perfumados* ("perfumers").

159. *Nostra Senyora de Consepsió* ("Our Lady of the Conception"). See the *Dietari*, vol. 15, 168–73 and 183 (July 17 and August 25, 1651, respectively). Despite bitter arguments among certain religious orders in the early seventeenth century regarding this aspect of Marian devotion, Barcelona had long favored the recognition of the Immaculate Conception. In fact, in the early 1640s the great hall of the Council of One Hundred in the City Hall was redecorated with a new program featuring this motif (Duran, *Barcelona*, I,

308). All the same, it is possible that Girona's vow to the Immaculada and St. Francis de Paul on August 15, 1650, served as the most immediate precedent for this action (Clara, "Pesta del 1650," 175).

160. Vilafranca del Penedès, a town thirty miles to the southwest of Barcelona.

161. These he called *dressanes*, or royal shipyards located in the port between the *Rambla* and the bastion of Saint Madrona in the southeast corner of the city.

162. In fact, the Council of One Hundred often found it hard to meet during the plague period, due to its inability to make a quorum. On April 13, 1651, the councillors noted "the difficulties the Council has in meeting because of the present sickness, as many members have left the city" (*Deliberacions*, vol. 160, 197v.). Thus on April 15 the council agreed to lower its quorum to forty-one members "from all estates" (ibid. 203r.–206r.).

163. Details about this painting are provided in *Datos Históricos*, 542; and Duran, *Barcelona*, II, 96–97 (the painting itself, now in the Cathedral Museum, is reproduced between pp. 104–5 of the latter).

164. A coastal town eighteen miles northeast of Barcelona.

165. Neither Holy Week nor Corpus Christi were celebrated in regular fashion in Barcelona in 1652 as a result of the plague and threat of siege (Ms. 225, 60r. and 63r.).

166. Parets's curious term for Irish mercenaries, *soldats ermànicos* ("Germanic soldiers"), first used in Ms. 225, 101v.

167. After the surrender of Barcelona in October 1652 and its reincorporation into the Spanish monarchy, Catalan resources were quickly mobilized for the war against the French. Details on the Girona campaign of 1653 can be found in Ms. 225, 100v., and Sanabre, *Acción de Francia*, 555–56.

168. The link between troop movements and the transmission of plague—frequently remarked in this period—can also be found in *Deliberacions*, vol. 161, 487r.–v. (the November 26, 1652, report of the Plague Board to the viceroy).

169. The Nazareth convent was the Barcelona proctorship, or house of the Cistercian monastery of Poblet, located some forty-five miles southwest of Barcelona (Madurell, "Priorat").

170. The viceroy Don Juan of Austria (1629–1679), also known as Juan José de Austria. Commander of the Spanish army which besieged Barcelona beginning in 1651, he was appointed viceroy and undertook a policy of conciliation toward the formerly rebellious Catalans.

171. The chapel of the old royal palace, on the King's Square.

172. This order earned a special reputation during the early modern period for the aid its members extended to the sick during the plague, as symbolized by the memorable figure of Fra Cristoforo in *I Promessi Sposi*. For a

brief account of the fate of local Capuchins during the Barcelona epidemic, see Rubí, *Segle*, 732–40.

173. Parets uses the term *cotreta*, synonym *quatreta* ("quartet"), or committee of four administrators or supervisers.

174. In original, *tallaven les cames al mal* ("they cut the plague's legs off").

175.. This sentence highlights the overtly contractual nature of therapeutic devotion during the early modern era. For comparisons elsewhere, see Christian, *Local Religion* and *Apparitions*.

176. The Tribunal of the Holy Office was housed in the viceregal palace (now the Archive of the Crown of Aragon) between the King's Square and the Cathedral.

177. Saints Philip and James the Lesser (May 1).

178. Santa Maria del Pi, or Saint Mary's of the Pine, the second largest parish in the city.

179. These were all typical figures of the Barcelona Corpus procession. The horses were not alive but figures like the rest and were traditionally the responsibility of the guild of *cotoners*, or cotton workers.

180. The *obrers*, or members of the vestry, were elected annually from a broad selection of parish members to oversee the maintenance accounts of the church, and to serve as advisers to the parish clergy on matters of finance and administration.

181. Parets mentioned this Capuchin preacher earlier in his text (Ms. 225, 83r.). Very few plague sermons survive from this era, even though some of their authors had been writers of considerable renown (such as the English poet John Donne, who preached a sermon at Saint Paul's in January 1626 on the recent epidemic in London). Sant Feliu (d. 1685) apparently left behind a written *Sermonari*, according to Rubí, *Segle*, 775, 780; however, I have not been able to consult this manuscript.

182. This was a Dominican convent near Saint Anne's Square (see Ainaud, *Catálogo Monumental* 165; Duran, "El monestir de Montsió," in *Barcelona*, I, 542–44).

183. This refers to the Dominican convent on the present-day Plaça dels Angels in the *Raval*, not to be confused with the older, extramural foundation which served as the first pesthouse in 1651.

184. This was one of the oldest monasteries in Barcelona. Previously located on the street of that name in front of the general hospital, it was demolished in 1874.

185. This is a commercial thoroughfare linking the *Rambla*, Barcelona's main thoroughfare, to the center of the city.

186. The site of the former Jewish ghetto of Barcelona, largely destroyed during the pogrom of 1391, located between the *Bocaria* and Saint James's Square.

187. Parets uses the Spanish word *palacio* to refer to the new royal

residence constructed by Don Juan of Austria to replace the older royal palace (and viceregal residence) between the King's Square and the Cathedral.

188. Saint Nicholas of Tolentino (1245–1305) was an Augustinian hermit and preacher from Macerata (Italy), canonized in 1446. Protector of souls in purgatory, his sign, the lily, symbolized penitence and purity. Frequently invoked against the plague (see Vaquer, "Peste de 1652," 136; Serrano, *Anacardina*, 9v.; Ballesteros, *Peste en Córdoba*, 118 for other seventeenth-century examples), by 1673 the Barcelona association of this name had merged with other confraternities sponsored by the Augustinians (A.C.A./ Monacals, Hisenda, 3306–3307, list of confraternity members 1673–1758). Significantly, the other Augustinian convent in Barcelona (the Discalced foundation of Saint Monica, where Parets's son Gabriel professed in 1664) also housed a chapel devoted to Sant Nicolau de Tolentí (Barraquer, *Casas de Religiosos*, II, 507).

189. A convent on Old Saint Augustine's Square that was demolished in the early eighteenth century to make room for the construction of the Ciutadella fortress (Ainaud, *Catálogo Monumental*, 170–73, and Martí i Bonet et al., *Convent i Parròquia*). Beginning on July 6, 1651, it was used as a plague hospital for "rich men," according to Bruniquer (*Rúbriques*, IV, 302) and *Ordinacions* (vol. 33, 18v.–20r., July 29–30, 1651).

190. A street linking the Sea Gate to the Wool Square, en route to the monastery of Saint Augustine's.

191. A street running behind the City Hall toward the sea to Broad Street.

192. Saint Mary's by the Sea.

Appendix I

1. Thomas Platter (ca. 1499–1582) grew up in a peasant family in the mountainous Valais region of southern Switzerland. He left his home while a teenager to begin a life as a wandering scholar and apprentice ropemaker, travelling as far east as Silesia, and alternating his trade with teaching in municipal schools and working in print shops. Platter eventually became one of the best-known humanist scholars and supporters of the Reformation in Switzerland. His *Lebensbeschreibung*—written as a book of advice and admonition to his son Felix, the only one of his children to have survived the family's numerous bouts with plague—is one of the most famous autobiographical accounts of the early modern period. I have translated the excerpts here from Marie Helmer's French version of the original German text (Platter, *Autobiographie*, 82–83, 88–91). Pages 129–30 of this edition contain a brief summary of the deaths the plague caused in Platter's family. Given the lack of chronological references within the text, I have not been able to ascertain the years in which the two episodes recounted here took place.

2. The cathedral of Zürich. The Protestant Reformer Ulrich Zwingli was appointed "people's priest" there in 1518.

3. Villages to the southeast of Bern, near Stalden, the area where Thomas was born and grew up.

4. Rudolf Gwalter (1519–1586), Protestant clergyman and Zwingli's son-in-law.

5. A town 25 miles southwest of Basel.

6. Thomas was in the service of Johannes Epiphanius, a physician and Protestant intellectual of Italian origin (*Autobiographie*, 86).

7. A coin circulating in Switzerland and southern Germany, worth one-fifth of a florin (*Autobiographie*, 34n.).

8. A village midway between Porrentruy and Basel.

9. *Welsches*, a derogatory term in German for speakers of Romance languages, usually applied to the French.

10. A village eight miles south of Delémont.

11. Virtually nothing is known of the author of this autobiographical relation of the Milanese plague of 1576 except that he was born in 1552 and lived in the *Borgo degli Ortolani*, a popular quarter located outside the Porta Comasine in the northwest part of the city. His trade is unknown, although the editor of the original manuscript (now in the Biblioteca Trivulziana, Milan) refers to him as a *popolano*, that is, of artisan background. The text was written in Lombard dialect and edited by "C.E.V." in 1877 (Cozzi, "Diario").

12. Certainly a curious way of referring to his own paternal grandfather, unless the reference to his "uncle" (*barba*) should not be taken literally.

13. Casale (1534–1629?), a carpenter by profession, left behind a sober mixture of personal autobiography and civic chronicle covering the years 1554 to 1598. His main purpose in doing so was to provide posterity with a detailed record of the achievements of the Archbishop of Milan, Saint Charles Borromeo, and in particular of his founding of the *Scuole di Dottrina*. These were a system of primary schools charged with imparting the rudiments of reading and writing along with Catholic doctrine to children from artisan families. Casale himself served as a teacher in these schools, along with his wife, two of their children, and other members of his exceptionally devout family. The original manuscript of his text, now in the Biblioteca Ambrosiana of Milan, was edited by Carlo Marcora in 1965 (Casale, "Diario"); pages 289 to 309 cover the Milanese plague of 1576, the same described by Giovan Ambrosio de' Cozzi. The excerpts translated here can be found on pages 290–92 and 297–300.

14. *Meregnano* in the original text, a town twenty miles southeast of Milan, site of the famous battle of Marignano between the French and the Swiss in 1515.

15. *crodavano come mosche* ("people dropped like flies").

16. The pesthouse, located outside Milan's eastern gate, was originally

established under the auspices of the city's General Hospital in 1488. Final work on the building was completed in 1629 under the supervision of Alessandro Tadino, superintendent of public health and author of a famous narrative of the plague of 1630 (Cozzi, "Diario," 129n.).

17. A street in the center of Milan, opening onto the Cathedral square.

18. *scuole de umanità*, or secondary schools, like the famed "Broletto" school, where in the mornings "lessons in humanities" were taught (*Ducato di Milano*, introduction).

19. The *Santo Chiodo* from the Crucifixion, one of Milan's most treasured relics, was housed in the Cathedral (Buratti, *Città rituale*, 115–19).

20. Saint Charles Borromeo served as archbishop of Milan from 1565 to 1584. See his *Memoriale* for a brief summary of his activity during the plague of 1576, along with the original text of his famous "Memoir to the Milanese," bitterly reproving the citizens' behavior during the epidemic.

21. Variant of the name Ginevra.

22. Casale concluded his account of the plague with a discussion of the measures ordered by the city government in late 1576 and 1577 (pp. 301–05), which was followed by an impersonal listing of the members of his family who died during the contagion (pp. 305–09).

23. Risi (1559?–1630) descended from a well-known family of glass and majolica-makers from Bologna. Socially prominent and well-to-do (the Risi owned several shops and industrial establishments), Risi also gathered an impressive collection of paintings, due in part to his personal friendship with contemporaries like Lodovico Carracci. The portions of his *Raccordi* (the original manuscript is in the Archivio di Stato of Bologna) referring to his art collecting were excerpted by Francesco Malaguzzi in an article published in 1918 (Malaguzzi, "Collezionista bolognese"). The reference translated here to the plague which killed Risi's children, wife, and ultimately himself can be found on page 46 of this article.

24. Serrano (d. 1657) was born in Salamanca and exercised the printer's trade in Madrid, Seville, Osuna, and Granada before settling in Málaga around 1630. During his lengthy publishing career, Serrano printed at least five books dealing with plague, the most detailed being his own treatise of 1650 entitled *Anacardina espiritual*. This he wrote expressly, as he notes on the title page, "to preserve the memory of the warnings that Divine Justice has sent to the city of Málaga." The most devastating of these warnings was the epidemic of 1649, the central theme of the book, from which the following excerpts derive (Serrano, *Anacardina*, 5r.–7r., 10v., and 13r.).

25. *regidores*, or city councillors, whose posts were normally transmitted by inheritance within a small handful of aristocratic families.

26. This measure was probably taken so as not to alarm the populace and to avoid broadcasting the fact that plague was in the city.

27. *berberiscos*, most probably slaves of African origin who were employed as stretcher bearers and grave diggers.

28. We know virtually nothing of the life of the author of this interesting chronicle of life in mid-seventeenth-century Seville, save that he owned a shop near the Cathedral on the Calle de Francos and that, having survived the epidemic, was still around to write a description of the popular revolt of 1652 known as the *motín de la Feria* (Vega, *Memorias*, 130–32). Curiously, his narrative resembles that of Parets in at least one respect: both authors resorted to writing in the first person only when reporting upon the plague. Vega's description of the epidemic of 1649 can be found on pages 120–28; a separate listing of religious processions during the plague (found elsewhere in his manuscript, located in the Biblioteca Capitular y Colombina of the Cathedral of Seville) is reproduced on pages 157–76 of the edition by Francisco Morales Padrón. I have translated passages from pages 120–25 and 176 of this edition.

29. Triana was (and still is) a popular extramural neighborhood on the left bank of the Guadalquivir River. The Street of the Virgins is located between the Minjoar Gate and the parish Church of San Nicolás, in the midst of one of the most heavily populated artisan districts of the city. The *Hierro Viejo* was probably the area known as the *Herrerías Reales*, or Royal Smithies, an impoverished commercial quarter fronting on the river near the Arenal Square. (I am indebted to Carlos Martínez Shaw for sharing with me his expert knowledge of the social topography of early modern Seville.)

30. *Torre de Oro*, famous medieval landmark near the river in the southern part of the old city.

31. "house . . . where many people lived" (*casa . . . de vecindad*). Also known as a *corral*, this multi-story, multi-family dwelling centering around an open courtyard remains a typical housing unit in southern Spanish cities to the present day.

32. Vega shifts here to the second person singular.

33. Born in the early seventeenth century in the village of Martín del Río in the nearby province of Teruel, Estiche died shortly after the plague of 1652. He apparently studied some medicine as well as surgery prior to his graduation from the University of Saragossa in 1650. Estiche wrote two books in his lifetime: a treatise on surgery, and a first-person account of the plague of 1652, from which the following passages have been translated (*Tratado*, 3r.–4r., 6v.–7r.). His description of the epidemic is obviously indebted to the treatise published a century earlier by the Sardinian physician Juan Tomás Porcell; in particular, both works contain detailed descriptions of the autopsies they performed on plague victims.

34. According to popular legend, the Apostle Santiago (Saint James the Great) was visited while preaching in Saragossa by a vision of the Virgin, who descended from heaven on a pillar, and ordered him to build a church on that

spot (now on the south bank of the Ebro River within the old city). The Virgin was declared patroness of Saragossa in 1642.

Appendix II

1. Needless to say, the list in this appendix is far from exhaustive, as there are hundreds of autobiographical accounts of the experience of plague. This brief repertory, moreover, is clearly biased toward southern Europe; further research would doubtless come up with many more accounts, especially from France, Germany, and the Low Countries.

Appendix III

1. *Dietari*, vol. 15, 157–58. *Datos Históricos*, 530–52, contains a Spanish translation of the same text. Interestingly, a remarkably similar lament was entered into the council minutes during the siege of 1652 (see the *Dietari*, vol. 15, 312–18, September 19, 1652).

2. The "careless" attitude and behavior of the bearers of the dead was a commonplace of plague chronicles. For example, the introduction to the First Day of the *Decameron* has Pampinea refer to the grave diggers' "insulting" and "vile songs." For one chronicle's report of sexual profanations of corpses by their bearers, see Carmichael, *Plague and the Poor*, 146.

3. *favar*.

4. "men cannot even remember themselves" (*ni los homens se recordan del ser que tenen*). I confess to finding this one of the most curious phrases in all the documents I have seen from early modern Barcelona. Equally extraordinary is this text's concern with the impact of plague on imagination and memory, especially its awareness of how the inadequacy of the former to conceive, much less express, the experience of plague shapes the public and collective articulations of the latter.

Appendix V

1. This popular antiquarian was one of the first scholars to consult the Parets chronicle, although it is possible that his near-contemporary Narcís Feliu de la Peña also used Parets for his account of the 1651 epidemic. However, Feliu explicitly cited the tanner's chronicle only for the years 1626–1640, which suggests that he had access only to its first volume (Ms. 224).

2. In a memorandum drawn up in 1736, Serra i Postius referred to his ownership of the following manuscript: "Anonymous, Barcelonan Catalan,

who wrote Spanish in a clear style, and concretely and realistically, as he had been an eyewitness of nearly everything he recounted: *History of the City of Barcelona During the Three Fatal Years 1650 1651 and 1652 During Which Time the City and Catalonia Suffered Hunger, Plague and War*. This manuscript was among the books of Serra i Postius in the year 1736" (B.N./Ms. 13.604, 106r.). The inventory of his books and manuscripts drawn up after his death in 1748 indicates that this document was still in his possession (Madurell, "Pedro Serra Postius," 382). This copy was donated in the later nineteenth century to the Biblioteca-Museu Victor Balaguer in Vilanova i la Geltrú, where it is presently. I am grateful to Miquel Angel Martinez and Teresa Besora for their help in obtaining a copy of this document.

3. *Memorias*, 683.

4. *Historia de Cataluña*, IV, 520–41.

5. The introduction to the M.H.E. edition was the only study of Parets he published, although in it he noted his intention to edit the original Catalan version of the chronicle. The M.H.E. version was consulted by the authors of the *Datos Históricos*, where it is referred to (without attribution) on page 496.

6. *Population catalane*, 26–37. Parets is also cited in Nadal, *Población*, 86–87. Both authors preferred the B.C./Ms. 502 version of the chronicle to the M.H.E. edition, in the understandable belief that the former was a more reliable version. Close comparison of the two translations with the Catalan original, however, reveals that the M.H.E. version, while filled with errors, is in general much more accurate.

7. "Peste en Cataluña," a summary of his original doctoral dissertation, which I have not been able to consult.

8. It is equally striking that the plague of 1651 itself finds so little mention in the more general histories of the period. For instance, the most detailed work on mid-seventeenth-century Catalonia—Sanabre's *Acción de Francia*—gives the epidemic only the most cursory attention (p. 486).

References

Advertiments Convenients per lo Govern Polítich de la Ciutat de Barcelona, en precautió de la pesta. Donats per lo Col.legi de Doctors en Medicina y de Chirurgia, a petitió dels molts Illustres senyors Consellers, y dotsena de morbo de dita Ciutat. Barcelona, 1652.

Ago, R., and A. Parmeggiani, "La peste del 1656–57 nel Lazio." *Actes I Congrés Hispano-Luso-Italià de Demografia Històrica*, 194–201. Barcelona, 1987.

Ainaud, J., J. Gudiol, and F.-P. Verrié, *Catálogo Monumental de España: La Ciudad de Barcelona*. Madrid, 1947.

Alberti, L. B. *The Family in Renaissance Florence*, edited by R. N. Watkins. Columbia, SC, 1969.

Alcanyis, Ll. "Regiment Preservatiu de la Pestilència." In *Butlletí de la Biblioteca de Catalunya*, edited by J. Rubió i Balaguer, (1923–1927): 32–57.

Alcover, A. M. *Diccionari català-valencià-balear*. 10 vols. Palma de Mallorca, 1930–1968.

Alexander, J. T. *Bubonic Plague in Early Modern Russia: Public Health and Urban Disaster*. Baltimore, 1980.

Amelang, J. S. "Honored Citizens and Shameful Poor: Social and Cultural Change in Barcelona, 1510–1714." Ph.D. diss. Princeton University, 1982.

———. *Honored Citizens of Barcelona: Patrician Culture and Class Relations, 1490–1714*. Princeton, 1986.

Andel, M. A. van. "Plague Regulations in the Netherlands." *Janus* 21 (1916): 410–44.

Anglada Vilardebó, J. "Del Testament en cas de Pesta." *Revista Jurídica de Catalunya* 73 (1974): 209–15.

Aragonés, D. *Lamentación y Duelo de la insigne ciudad de Barcelona, sobre el estrago que le ha causado el pestífero Morbo en el año 1589*. Barcelona, 1590.

Arditi, B. *Diario di Firenze e di altre parti della Cristianità, 1574–1579*. Edited by R. Cantagalli. Florence, 1970.

Baehrel, R. "Épidémies et terreur: Histoire et Sociologie." *Annales Historiques de la Révolution Française* 23 (1951): 113–46.

———. "La haine de classe en temps de l'épidémie." *Annales E.S.C.* 7 (1952): 351–60.

Balaguer, V. *Historia de Cataluña y de la Corona de Aragón.* 5 vols. Barcelona, 1860–1863.

———. *Las Calles de Barcelona.* 2 vols. Barcelona, 1866.

Ballesteros Rodríquez, J. *La Peste en Córdoba.* Córdoba, 1982.

Banker, J. "Mourning a Son: Childhood and Parental Care in the *Consolatoria* of Giannozzo Manetti." *History of Childhood Quarterly* 3 (1976): 351–62.

Barraquer i Roviralta, C. *Las Casas de Religiosos en Cataluña durante el primer tercio del S. XIX.* 2 vols. Barcelona, 1906.

Barrett, W. P., ed. *Present Remedies against the Plague, etc.* Oxford, 1933.

Bastian, F. "Defoe's *Journal of the Plague Year* Reconsidered." *Review of English Studies,* n.s., 16 (1965): 151–73.

Baxter, R. *The Autobiography of Richard Baxter.* Edited by N. H. Keeble and J. M. Lloyd Thomas. London, 1974.

Bennassar, B. *Recherches sur les grandes épidémies dans le nord de l'Espagne à la fin du seizième siècle. Problèmes de documentation et de méthode.* Paris, 1969.

———. "L'Organisation municipale et communautés d'habitants en temps de peste: l'Exemple du nord de l'Espagne et de la Castille à la fin du seizième siècle." *Villes de l'Europe méditerranéenne et de l'Europe occidentale du moyen âge au 19e siècle,* 139–44. Nice, 1969.

Bertelli, S. *Rebeldes, Libertinos y Ortodoxos en el Barroco.* Barcelona, 1984.

Biraben, J.-N. *Les Hommes et la peste en France et dans les pays européens et méditerranéens.* 2 vols. Paris, 1975.

Boccaccio, G. *The Decameron.* Translated and edited by G. H. McWilliam. Harmondsworth, 1972.

Borromeo, C. *Memoriale ai Milanesi.* Edited by G. Testori and G. Pozzi Bellini. Milan, 1965.

Bowsky, W. M., ed. *The Black Death: A Turning Point in History?* Huntingdon, NY, 1978.

Brighetti, A. *Bologna e la Peste del 1630.* Bologna, 1968.

Brown, J. C. *In the Shadow of Florence: Provincial Society in Renaissance Pescia.* New York, 1982.

Brozzi, M. *Peste, Fede e Sanità in una Cronaca Cividalese del 1598.* Milan, 1982.

Brucker, G. *The Society of Renaissance Florence: A Documentary Study.* New York, 1971.

Brumont, F. "Le Coup de grâce: La Peste de 1599." *Actas del Congreso de Historia de Burgos*, 335–42. Burgos, 1984.

Bruniquer, E. G. *Rúbriques. . . . Ceremonial dels Magnìfics Consellers y Regiment de la Ciutat de Barcelona.* 5 vols. Barcelona, 1912–1916.

Buratti, A., G. Coppola, G. Crespi, P. Falzone, G. Mezzanotte, G. Oneto, G. B. Sannazzaro, and M. C. Verga. *La Città rituale: la città e lo stato di Milano nell'età dei Borromeo.* Milan, 1982.

Burgos, A. de. *Tratado de peste, su esencia, prevención y curación, con observaciones muy particulares.* Córdoba, 1651.

Burke, P. *The Historical Anthropology of Early Modern Italy: Essays on Perception and Communication.* Cambridge, 1987.

Busquets, J. "Població i societat a la Girona del S. 17. El testimoni de Jeroni de Real." In *Girona a l'Epoca Moderna: Demografia i Economia*, edited by R. Alberch, 85–106. Girona, 1982.

Caldera de Heredia, G. *Tribunal medicum, magicum et politicum. Pars III. Tractatus . . . de peste quae anno 1649 hispalensem civitatem . . . infecerat.* Leiden, 1658.

Calvi, G. "L'Oro, il fuoco, le forche: la peste napoletana del 1656." *Archivio Storico Italiano* 139 (1981): 405–58.

———. *Storie di un Anno di Peste: Comportamenti sociali e Immaginario nella Firenze Barocca.* Milan, 1984.

———. "A proposito di *Storie di un anno di peste.*" *Quaderni Storici* 21 (1986); 1009–18.

———. "'Dall'altrui communicatione': comportamenti sociali in tempo di peste." *Actes I Congrés Hispano-Luso-Italià de Demografia Històrica*, 184–93. Barcelona, 1987.

Camporesi, P. "Cultura popolare e cultura d'elite fra Medioevo e età moderna." *Storia d'Italia. Annali IV*, 81–157. Turin, 1981.

———. *Bread of Dreams: Food and Fantasy in Early Modern Europe.* Chicago, 1989.

Camps i Surroca, M., and M. Camps i Clemente. *La Pesta de meitats del S. XVII a Catalunya.* Lleida, 1985.

Canal i Morell, J. *Una Vila Catalana davant la Mort: La Pesta de 1650 a Olot.* Olot, 1987.

Capmany, A. de. *Memorias Históricas sobre la Antigua Ciudad de Barcelona.* 3 vols. Edited by E. Giralt and C. Batlle. Barcelona, 1961–1963.

Cardús, S. *Terrassa durant la Guerra dels Segadors.* Terrassa, 1971.

Carmichael, A. G. *Plague and the Poor in Renaissance Florence.* Cambridge and New York, 1986.

Carpentier, E. *Une Ville devant la peste: Orvieto et la peste noire de 1348.* Paris, 1962.

Carreras Panchón, A. "Dos Testimonios sobre la Epidemia de Peste de 1599 en Valladolid." *Asclepio* 25 (1973): 351–57.

———. *El Médico y la Peste en la España del Renacimiento*. Salamanca, 1976.

———. "Las Epidemias de Peste en la España del Renacimiento." *Asclepio* 29 (1977): 5–15.

Carreras Roca, M. "La Peste en Cataluña durante el S. XVII." *Medicina e Historia* 29 (1967): 9–19.

Casale, G. "Diario. . . . 1554–1598." Edited by C. Marcora. *Memorie storiche della diocesi di Milano* 12 (1965): 209–437.

Casey, J. "La Crisi general del S. XVII a València, 1646–48." *Boletín de la Sociedad Castellonense de Cultura* 46, (1970): 96–173.

Castells i Calzada, N. "La peste de mediados del S. XVII en Catalunya." *Actes I Congrés Hispano-Luso-Italià de Demografia Històrica*, 104–12. Barcelona, 1987.

Cerutti, S. "Matrimoni del tempo di peste: Torino nel 1630." *Quaderni Storici* 55 (1984): 65–106.

Céspedes y Meneses, G. de. *Historias Peregrinas y Ejemplares*. Edited by Y. R. Fonquerne. Madrid, 1969.

[Chamberlain, J.] *The Chamberlain Letters: A Selection of the Letters of John Chamberlain Concerning Life in England from 1597 to 1626*. Edited by E. Thomson. New York, 1966.

Charlier, J. *La Peste à Bruxelles de 1667 à 1669 et ses conséquences démographiques*. Brussels, 1969.

Chartier, R. "Culture as Appropriation: Popular Cultural Uses in Early Modern France." In *Understanding Popular Culture: Europe from the Middle Ages to the Nineteenth Century*. Edited by S. L. Kaplan, 229–54. Berlin, 1984.

Christian, W. A. *Local Religion in Sixteenth-Century Spain*, Princeton, 1981.

———. *Apparitions in Late Medieval and Renaissance Spain*. Princeton, 1981.

Cipolla, C. M. *Faith, Reason and the Plague in Seventeenth-Century Tuscany*. New York, 1981.

———. *Contro un Nemico Invisibile: Epidemie e Strutture Sanitarie nell'Italia del Rinascimento*. Bologna, 1985.

Clara, J. "La Pesta del 1650. La Desigualtat devant la Mort i Aspectes religioses." In *Girona a l'Epoca Moderna: Demografia i Economia*, edited by R. Alberch et al., 165–88. Girona, 1982.

Clark, S. "French Historians and Early Modern Popular Culture." *Past and Present* 100 (1983): 62–99.

Cochrane, E. *Historians and Historiography in the Italian Renaissance.* Chicago, 1981.

[Cozzi, G. A. de'.] "Diario di un popolano milanese durante la peste di 1576." Edited by C.E.V. *Archivio Storico Lombardo* 4 (1877): 124–40.

Crawfurd, R. *Plague and Pestilence in Literature and Art.* Oxford, 1914.

"Curiosi Ricordi del Contagio di Firenze del 1630." Edited by D. Catellacci. *Archivio Storico Italiano* 20 (1897): 379–91.

Cuvelier, E. "*A Treatise of the Plague* de Thomas Lodge: Traduction d'un ouvrage médicale français." *Études Anglaises* 21 (1968): 395–403.

D'Amelia, M. "La peste romana del 1656–57 e l'intervento dell'annona." *Actes I Congrés Hispano-Luso-Italià de Demografia Històrica,* 202–14. Barcelona, 1987.

David-Peyre, Y. "La Peste et le mal vénérien dans la littérature ibérique du 16e et du 17e siècle." Ph.D. diss., University of Poitiers, 1967.

———. *Le Personnage du médecin et la relation médecin-malade dans la littérature ibérique, 16–17e S.* Paris, 1971.

Defoe, D. *A Journal of the Plague Year.* Edited by J. H. Plumb. New York, 1960.

Delumeau, J. *La Peur en occident, 14e–18e siècles: Une Cité assiégée.* Paris, 1978.

Deyon, P. "Mentalités populaires: un sondage à Amiens au 17e siècle." *Annales E.S.C.* 17 (1962): 448–58.

Dols, M. "Comparative Communal Responses to the Black Death in Muslim and Christian Societies." *Viator* 5 (1974): 269–87.

Domínguez Ortiz, A. *Sociedad y Mentalidad en la Sevilla del Antiguo Régimen,* rev. ed. Seville, 1983.

Il Ducato di Milano in età spagnola, 1535–1713: Architettura, urbanistica e assetto del territorio. Milan, 1990.

Duran i Sanpere, A. *Barcelona i la seva història.* 3 vols. Barcelona, 1973.

Dyer, R. D. "The Influence of Bubonic Plague in England, 1500–1667." *Medical History* 22 (1978): 308–26.

Elliott, J. *The Revolt of the Catalans: A Study in the Decline of Spain, 1598–1640.* Cambridge, 1963.

Emery, R. W. "The Black Death of 1348 in Perpignan." *Speculum,* 42 (1967): 611–23.

Erikson, K. T. *In the Wake of the Flood.* London, 1979.

Estiche, J. *Tratado de la Peste de Çaragoça en el año 1652.* Pamplona, 1655.

L'Étoile, P. de. *The Paris of Henry of Navarre. . . . Selections from his*

Mémoires-Journaux. Edited by N. Lyman Roelker. Cambridge, MA, 1958.

Evelyn, J. *The Diary of John Evelyn*. Edited by J. Bowle. Oxford, 1985.

Fatjó, P. "Un instrument de la caritat eclesiàstica: la Pia Almoina." *L'Avenç* 91 (March 1986): 44–47.

Feliu de la Peña, N. *Anales de Cataluña*. 3 vols. Barcelona, 1709.

Ferran, J., F. Viñas y Cusí, and R. de Grau, "Datos históricos sobre las epidemias de peste ocurridas en Barcelona." In their *La Peste Bubónica. Memoria sobre la Epidemia ocurrida en Porto en 1899*, 369–625. Barcelona, 1907.

Fornés, J. *Tractatus de peste, praecipue gallo-provinciali, et occitania grassanti, in quinque partes divisus*. Barcelona, 1725.

Fortea Pérez, J. I. *Córdoba en el S. XVI: Las Bases Demográficas y Económicas de una Expansión Urbana*. Córdoba, 1981.

Friedman, E. G. "Public Health in Seventeenth-Century Catalonia: A Conflict over Jurisdiction." *Actes del Primer Congrés d'Història Moderna de Catalunya*, I, 581–86. Barcelona, 1984.

Gaffarel, P., and M. de Duranty. *La Peste de 1720 à Marseille*. Paris, 1911.

Galera, M., F. Roca, and S. Tarragó. *Atlas de Barcelona, segles XVI–XX*. Barcelona, 1982.

Galpern, A. N. *The Religions of the People in Sixteenth-Century Champagne*. Cambridge, MA, 1976.

García Ballester, L., and J. M. Mayer Benítez. "Aproximación a la historia social de la peste de Orihuela de 1648." *Medicina Española* 65 (1971): 317–33.

García de Enterría, M. C. *Sociedad y poesía de cordel en el Barroco*. Madrid, 1973.

Gavaldá, F. *Memoria de los Sucessos Particulares de Valencia y su Reino en los años mil seiscientos quarenta y siete y quarenta y ocho, tiempo de peste*. Valencia, 1651. Reprint edited by M. Peset. Valencia, 1979.

Ginzburg, C. *The Cheese and the Worms: The Cosmos of a Sixteenth-Century Miller*. Harmondsworth, 1982.

———. "The Dovecote Has Opened Its Eyes: Popular Conspiracy in Seventeenth-Century Italy." In *The Inquisition in Early Modern Europe: Studies on Sources and Methods*, edited by G. Henningsen and J. Tedeschi, 190–98. Dekalb, IL, 1986.

Giralt, E. "La colonia mercantil francesa de Barcelona a mediados del siglo XVII." *Estudios de Historia Moderna* 6 (1956): 215–78.

Glückel of Hameln. *Memoirs*. Edited by Marvin Lowenthal. New York, 1977.

Granjel, L. S. "Las Epidemias de Peste en la España del Siglo XVII." *Asclepio* 29 (1977): 17–36.

———. *La Medicina española del S. XVII.* Salamanca, 1978.
———. *La Medicina española renacentista.* Salamanca, 1980.
Grendi, E. "Storia sociale e storia interpretativa." *Quaderni Storici* 21, (1986): 201–10.
Greyerz, K. von. "Religion in the Life of German and Swiss Autobiographers (Sixteenth and Early Seventeenth Centuries)." In *Religion and Society in Early Modern Europe, 1500–1800*, edited by K. von Greyerz, 223–41. London, 1984.
Griffin, C. *The Crombergers of Seville: The History of a Printing and Merchant Dynasty.* Oxford, 1988.
Guerchberg, S. "La controverse sur les prétendus semeurs de la 'Peste Noire' d'après les traités de peste de l'époque." *Revue des Études Juives*, n.s., 8 (1948): 3–40.
La Guerra e la Peste nella Milano dei "Promessi Sposi." Documenti inediti tratti dagli archivi spagnoli. [Edited by C. Greppi.] Madrid, 1975.
Guicciardini, F. *The History of Italy.* Edited by S. Alexander. Princeton, 1982.
Guío Cerezo, Y. "El influjo de la luna: acerca de la salud y la enfermedad en dos pueblos extremeños." *Asclepio*, 40 (1988): 317–41.
Haton, C. *Mémoires de Claude Haton contenant le récit des événements accomplis de 1553 à 1582, principalement dans la Champagne et la Brie.* 2 vols. Edited by F. Bourquelot, Paris, 1857.
Henderson, J. Review of Ann G. Carmichael, *Plague and the Poor in Renaissance Florence.* [The London] *Times Literary Supplement*, February 20, 1987, p. 189.
Herlihy, D. *Medieval Households.* Cambridge, MA, 1985.
Hildesheimer, F. "Prévention de la peste et attitudes mentales en France au 18e siècle." *Revue Historique* 265 (1981): 65–79.
Hill, C. "The Medical Profession and Its Radical Critics." In his *Change and Continuity in Seventeenth-Century England*, 157–78. Cambridge, MA, 1975.
Hirst, L. F. *The Conquest of Plague: A Study of the Evolution of Epidemiology.* Oxford, 1953.
Iglésies, J. *Pere Gil, S.I. (1551–1622) i la seva Geografia de Catalunya.* Barcelona, 1949.
Kent, D. *The Rise of the Medici: Faction in Florence, 1426–1434.* Oxford, 1978.
King, M. L. "An Inconsolable Father and His Humanist Consolers: Jacopo Antonio Marcello, Venetian Nobleman, Patron, and Man of Letters." In *Iter Festivum: A Festschrift for Paul O.*

Kristeller, edited by J. Hankins, J. Monfasani, and F. J. Purnell, Jr. Binghamton, NY, 1986.

Klebs, A. C., and E. Droz. *Remèdes contre la peste. Facsimiles, notes et liste bibliographique des incunables sur la peste*. Paris, 1925.

Langa, M. de. *Relación muy verdadera, de las crueldades e imposiciones del Conde Duque en toda la monarquia de España, y particularmente la depravada voluntad con que ha deseado destruyr, y aniquilar el Principado de Cataluña y ciudad de Barcelona, compuesta por Martín de Langa ciego*. Barcelona, 1641.

[Lantery, Raymond de.] *Un Comerciante Saboyano en el Cádiz de Carlos II: Las Memorias de Raimundo de Lantery, 1673–1700*. Edited by M. Bustos Rodríguez. Cádiz, 1983.

La Parra, S. *Tiempo de Peste en Gandía, 1648–1652*. Gandía, 1984.

Lebrun, F. *Les Hommes et la mort en Anjou aux 17e et 18e siècles. Essai de démographie et psychologie historiques*. Paris, 1975.

León, P. de. *Grandeza y Miseria en Andalucía: Testimonio de una Encrucijada histórica, 1578–1616*. Edited by P. Herrera Puga. Granada, 1981.

Le Roy Ladurie, E. "A Concept: The Unification of the Globe by Disease (Fourteenth to Seventeenth Centuries)." In his *The Mind and Method of the Historian*, 28–83. Chicago, 1984.

Lombardi, D. "1629–1631: crisi e peste a Firenze." *Archivio Storico Italiano* 137 (1979): 3–50.

López de Meneses, A. "Una consecuencia de la peste negra en Cataluña: el pogrom de 1348." *Sefarad* 19 (1959): 92–131.

López Ferreiro, A. *Historia de la Santa A.M. Iglesia de Santiago de Compostela*. vol. 8. Santiago, 1905.

López Piñero, J. M. and F. M. Terrada. "La Obra de Porcell y los Orígenes de la Anatomía Patologíca Moderna." *Medicina e Historia* 34 (1967): 1–15.

López Piñero, J. M., T. F. Glick, V. Navarro Brotóns, and E. Portela Marco. *Diccionario histórico de la ciencia moderna en España*. 2 vols. Barcelona, 1983.

Lottin, A. *Chavatte, ouvrier Lillois. Un Contemporain de Louis XIV*. Paris, 1979.

Loyola, Ignatius of. *Inigo: Discernment Log-Book. The Spiritual Diary of St. Ignatius Loyola*. Edited by J. A. Munitiz. London, 1987.

Macfarlane, A. "*The Family, Sex and Marriage in England 1500–1800* by Lawrence Stone." *History and Theory* 18 (1979): 103–26.

Madurell i Marimon, J. M. "Pedro Serra Postius." *Analecta Sacra Tarraconensia* 29 (1957): 345–400.

———. "El priorat de Santa Maria de Natzaret de Barcelona, 1311–1660." *Miscellània Pobletana*, 267–83. Poblet, 1966.

Malaguzzi Valeri, F. "Un collezionista bolognese del Seicento." *L'Archigimnasio* 13 (1918): 29–48.
Manzoni, A. *I Promessi Sposi*. Edited by M. Barbi and F. Ghisalberti. Milan, 1942.
———. *Storia della colonna infame*. Milan, 1842.
Martí i Bonet, J. M., J. M. Juncà i Ramon, and L. Bonet i Armengol. *El Convent i Parròquia de Sant Agustí de Barcelona*. Barcelona, 1980.
Martin, A. L. *The Jesuit Mind: The Mentality of an Elite in Early Modern France*. Ithaca, NY, 1988.
Martínez de Pisón, E. *Segovia: Evolución de un paisaje urbano*. Madrid, 1976.
Mas, B. *Orde Breu, y Regiment molt útil, y profitós per a Preservar, y Curar de Peste*. Barcelona, 1625.
McClure, G. W. "The Art of Mourning: Autobiographical Writings on the Loss of a Son in Italian Humanist Thought, 1400–1461." *Renaissance Quarterly* 29 (1986): 440–75.
Modena, L. *The Autobiography of a Seventeenth-Century Venetian Rabbi: Leon Modena's Life of Judah*. Edited by M. R. Cohen. Princeton, 1988.
Mollaret, H. H., and J. Brossollet. "La peste, source méconnue d'inspiration artistique." *Annuaire du Musée Royal des Beaux-Arts d'Anvers* (1965): 3–112.
Montaigne, M. de. *The Autobiography of Michel de Montaigne*. Edited by M. Lowenthal. New York, 1956.
Montemayor, J. "Una ciudad frente a la peste: Toledo a fines del XVI." In *La Ciudad Hispánica durante los S. XIII al XVI*. Vol. II, 1113–31. Edited by E. Sáez, C. Segura, and M. Cantera. Madrid, 1985.
Monter, E. W. *Witchcraft in France and Switzerland: The Borderlands during the Reformation*. Ithaca, NY, 1976.
Muir, E. *Civic Ritual in Renaissance Venice*. Princeton, 1981.
Mullett, C. F. *The Bubonic Plague and England*. Lexington, KY, 1956.
Nadal, J. "L'última pandèmia de pesta a Catalunya, 1650–1654." *II Congrés Internacional d'Història de la Medicina Catalana, 19–38*. Barcelona, 1977.
———. *La Población española, S. XVI–XX*, rev. ed. Barcelona, 1984.
Nadal, J., and E. Giralt. *La Population catalane de 1553 à 1717: L'Immigration française*. Paris, 1960.
Nadi, G. *Diario bolognese*. Edited by C. Ricci and A. Bacchi Della Lega. Bologna, 1886.
Nicholson, W. *The Historical Sources of Defoe's Journal of the Plague Year*. Port Washington, NY, 1966. [Reprint of original 1920 edition.]

Nicolini, F. *Peste e Untori nei* Promessi Sposi *e nella realtà storica*. Bari, 1937.

———. "La Peste del 1629–1632." In *Storia di Milano*. Vol. 10, 499–560. Milan, s.d.

Nussdorfer, L. "Civic Salvation under Absolutism: The Plague Threat of 1629–32." Chapter 6 of Ph.D. *City Politics in Baroque Rome, 1632–1644*. Princeton University, 1985.

Oliver-Smith, T. *The Martyred City: Death and Rebirth in the Andes*. Albuquerque, NM, 1986.

Parets, M. *Dietari d'un any de pesta*. Edited by J. S. Amelang and X. Torres. Vic, 1989.

Parfitt, G. ed. *Ben Jonson: The Complete Poems*. New Haven, 1975.

Parry, A. "The Language of Thucydides' Description of the Plague." *Bulletin of the Institute of Classical Studies* 16 (1969): 106–18.

[Pascual, P.] "Les mémoires du notaire Pierre Pascual." Edited by P. Masnou. *Revue d'histoire et d'archéologie du Roussillon*, 6 (1905): 178–192, 212–23, 245–56, 277–88, 309–20, 340–52.

Pastore, A. "Peste e società." *Studi Storici* 20 (1979): 857–73.

———. "Testamenti in tempo di peste: la pratica notarile a Bologna nel 1630." *Società e storia* 16 (1982): 263–97.

———. "Tra Giustizia e Politica: Il Governo della peste a Genova e Roma nel 1656–1657." *Actes I Congrés Hispano-Luso-Italià de Demografia Històrica*, 226–39. Barcelona, 1987. [Also published in slightly revised form in *Rivista Storica Italiana* 100 (1988): 126–54.]

Pepys, S. *Diary*. Vol. VI. Edited by R. Latham and W. Matthews. London, 1972.

Pérez Moreda, V. *Las Crisis de Mortalidad en la España Interior, S. XVI–XIX*. Madrid, 1980.

Pérouas, L. "L'Univers mental d'un magistrat marchois: Pierre Robert du Dorat (1589–1658)." *Actes 102e Congrés national des Sociétés Savantes, Limoges, 1977*, vol. II, 23–39. Paris, 1978.

Peset, J. L. "Los Médicos y la Peste de Valencia de 1647–1648." *Asclepio*, 29 (1977): 217–41.

Peset, M., and S. La Parra, "La demografía de la peste en Valencia de 1647–1648." *Asclepio*, 27 (1975): 197–231.

Pladevall i Font, A., and A. Simon i Tarrés, eds. *Guerra i Vida pagesa a la Catalunya del Segle XVII*. Barcelona, 1986.

Platter, T. *Autobiographie*. Edited by M. Helmer. Paris, 1964.

Pollitzer, R. *Plague*. Geneva, 1954.

Pollock, L. *Forgotten Children: Parent-Child Relations from 1500 to 1900*. Cambridge, 1983.

Poole, J. F. C., and A. J. Holladay, "Thucydides and the Plague of Athens." *Classical Quarterly* 73 (1979): 282–300.
Porcar, P. J. *Coses Evengudes en la Ciutat i Regne de València: Dietari, 1589–1628*. Edited by F. Garcia Garcia. Valencia, 1983.
Porcell, J. T. *Información y curación de la peste en Zaragoza, y preservación contra peste en general*. Saragossa, 1565.
Preto, P. *Peste e Società a Venezia nel 1576*. Vicenza, 1978.
[Prion, P.], *Pierre Prion, Scribe*. Edited by E. Le Roy Ladurie and O. Ranum, Paris, 1985.
Procopius. *History of the Wars I*. Edited by H. B. Dewing. London, 1914.
Pujades, J. *Dietari*. 4 vols. Edited by J. M. Casas Homs. Barcelona, 1975–1976.
Pullan, B. *Rich and Poor in Renaissance Venice: The Social Institutions of a Catholic State, to 1620*. Cambridge, MA, 1971.
Reher, D.-S. "Les ciutats i les crisis a l'Espanya moderna." *Estudis d'Història Agrària* 5 (1985): 91–114.
"Relazione della pestilenza accaduta in Napoli nel 1656." Edited by G. De Blasiis. *Archivio storico per le provincie napoletane* I (1876): 323–57.
Renzi, S. de. *Napoli nell'Anno 1656, ovvero Documenti della Pestilenza che desolò Napoli nel 1656 preceduti dalla Storia di quella tremenda sventura*. Naples, 1867. [facsimile ed., Naples, 1968]
Riera, J., and J. M. Jiménez Muñóz. "Avisos en España por la peste de Milán." *Asclepio* 25 (1973): 165–72.
———. "El Doctor Rossell y los Temores en España por la Peste de Milán, 1629–1631." *Asclepio* 29 (1977): 283–307.
Riggs, D. *Ben Jonson: A Life*. Cambridge, MA, 1989.
Ripamonti, G. *La Peste di Milano del 1630, Libri Cinque*. Edited by F. Cusani. Milan, 1841.
Roberts, R. S. "A Note on Thomas Lodge's *A Treatise of the Plague* (1603)." *Medical History* 22 (1978): 89.
Roiz Soares, P. *Memorial*. Edited by M. Lopes de Almeida. Coimbra, 1953.
Rojas Villandrando. A. de. *El Buen Repúblico*. Salamanca, 1611.
Rossell, J. F. *El Verdadero Conocimiento de la Peste, sus Causas, Señales, Preservación y Curación*. Barcelona, 1632.
Rowse, A. L. *Sex and Society in Shakespeare's Age: Simon Forman the Astrologer*. New York, 1974.
Rubí, B. de. *Un segle de vida caputxina a Catalunya, 1564–1664: aproximació històrico-bibliogràfica*. Barcelona, 1977.
Rubino, A. "Anno 1656. Peste crudele in Napoli." *Archivio storico per le provincie napoletane* 19 (1894): 696–710.

Sabuco de Nantes y Barrera, O. *Nueva Filosofía de la naturaleza del hombre*. Edited by A. Martínez Tomé. Madrid, 1981.

Sanabre, J. *La Acción de Francia en Cataluña, 1640–1659 en la Pugna por la Hegemonía de Europa*. Barcelona, 1956.

Sanz Sampelayo, J., "La epidemia de mediados del siglo XVII en Andalucía (1647–1650). Historiografía actual, aportaciones y nuevas notas." *Actes I Congrés Hispano-Luso-Italià de Demografia Històrica*, 155–63. Barcelona, 1987.

Sardi Bucci, D. "La peste del 1630 a Firenze." *Ricerche Storiche* 10 (1980): 49–92.

Schama, S. *The Embarrassment of Riches: An Interpretation of Dutch Culture in the Golden Age*. Berkeley, 1988.

Serra i Postius, P. "Lo Perqué de Barcelona." Edited by R. D. Perés. *Memorias de la Real Academia de Buenas Letras de Barcelona* 9 (IV) (1929).

Serrano de Vargas, J. *Anacardina espiritual (1650)*. Edited by A. Llordén. Málaga, 1957.

Shrewsbury, J. F. D. *A History of Bubonic Plague in the British Isles*. Cambridge, 1970.

Simón Díaz, J. "Otro Romance sobre Desgracias Logroñesas." *Berceo* 23 (1952): 243–52.

Singer, D. W. "Some Plague Tractates (14th and 15th Centuries.)" *Proceedings of the Royal Society of Medicine. Section of the History of Medicine* 9 (15 March 1916): 159–212.

Slack, P. *The Impact of Plague in Tudor and Stuart England*. Boston, 1986.

———. "Metropolitan Government in Crisis: The Response to the Plague." In *London 1500–1700: The Making of the Metropolis*, 60–81, A. L. Beier and R. Finlay, eds. London, 1986.

Smith, R. S. "Barcelona Bills of Mortality and Population, 1457–1590." *Journal of Political Economy* 44 (1936): 84–93.

Soria, J. *Dietari*. Edited by F. de P. Momblanch Gonzalbez. Valencia, 1960.

Swanson, H. "The Illusion of Economic Structure: Craft Guilds in Late Medieval English Towns." *Past and Present* 121 (1988): 29–48.

Tenenti, A. *Il Senso della Morte e l'Amore della vita nel Rinascimento*. Turin, 1957.

Thomas, K. *Religion and the Decline of Magic*. New York, 1971.

Thompson, E. P. "Happy Families." *Radical History Review* 20 (1979): 42–50.

Thucydides. *The Peloponnesian War*. Translated by R. Warner, with an Introduction and Notes by M. I. Finley. Harmondsworth, 1972.

Torreilles, P. "Mémoires d'un chirurgien au 17e siècle." *Revue d'histoire et d'archéologie du Roussillon* 4 (1903): 167–82, 199–215.

Torres Amat, F. *Memorias para ayudar a formar un Diccionario Crítico de los Escritores Catalanes*. Barcelona, 1836.

Torres Sánchez, R. "Expansión de la epidemia de 1648 en la región murciana." *Actes I Congrés Hispano-Luso-Italià de Demografia Històrica*, 121–27. Barcelona, 1987.

Trexler, R. "Measures against Water Pollution in Fifteenth-Century Florence." *Viator* 5 (1974): 455–68.

———. "In Search of Father: The Experience of Abandonment in the Recollections of Giovanni di Pagolo Morelli." *History of Childhood Quarterly* 3 (1975): 225–52.

———. *Public Life in Renaissance Florence*. New York, 1980.

Vaquer Bennasar, O. "La peste de 1652 en Mallorca." *Actes I Congrés Hispano-Luso-Italià de Demografia Històrica*, 128–36. Barcelona, 1987.

[Vega, A. de la.] *Memorias de Sevilla, 1600–1678*. Edited by F. Morales Padrón. Córdoba, 1981.

Venezia e la Peste, 1348–1797. Venice, 1979.

Veny i Clar, J., ed. "*Regiment de Preservació de Pestilèncias*" *de Jacme d'Agramont, S. XIV*. Tarragona, 1971.

Vilar, P. *La Catalogne dans l'Espagne moderne*. 3 vols. Paris, 1962.

Viñas i Cusí, F. *La Glànola a Barcelona. Estudi d'una de les Epidèmies, 1651–1654. Acció de l'Administració pública en l'Antigó, en lo Present y en lo Pervindre*. Barcelona, 1901.

Vincent, B. *Andalucía en la Edad Moderna: Economía y Sociedad*. Granada, 1985.

Vincentz, C. R., ed. *Die Goldschmiede-Chronik: Die Erlebnisse der ehrbaren Goldschmiede-Ältesten Martin und Wolfgang, auch Mag. Peters Vincentz*. Hanover, 1918.

Vovelle, M. *Mourir autrefois: Attitudes collectives devant la mort aux 17e et 18e siècles*. Paris, 1974.

———. "Les intermédiaires culturels." In *Idéologies et Mentalités*, 163–76. Paris, 1982.

Wigglesworth, M. *The Diary of Michael Wigglesworth, 1653–1657: The Conscience of a Puritan*. Edited by E. S. Morgan. New York, 1965.

Willen, D. "Women in the Public Sphere in Early Modern England: The Case of the Urban Working Poor." *Sixteenth Century Journal* 19 (1988): 559–75.

Wilson, F. P. *The Plague in Shakespeare's London*, rev. ed. Oxford, 1963.

Ziegler, P. *The Black Death*. New York, 1971.

Index